D1240435

CONTENTS

CONTRIBUTORS

Wolf R. Dombrowsky is Co-Director of the Katastrophenforschungsstelle, KFS at Christian Albrechts University, Olshausenstrasse 40, Kiel D-24098, Germany.

Russell R. Dynes is Research Professor of the Department of Sociology and Disaster Research Center at the University of Delaware, Newark, Delaware 19716, USA.

Claude Gilbert is Research Director of the Program on Collective Risks and Crisis Situations at the Centre National de la Recherche Scientifique, 2 rue General Marchand, 38000 Grenoble, France.

Valerie J. Gunter is an Assistant Professor in the Department of Sociology and Associate Director of the Environmental Social Science Research Institute, 356 Liberal Arts, at the University of New Orleans, New Orleans, Louisiana 70148, USA.

Kenneth Hewitt is Professor of the Department of Geography at Wilfrid University, Waterloo, Ontario, Canada N2L 3C5.

Gary A. Kreps is Professor of the Department of Sociology at the College of William and Mary, Williamsburg, Virginia 23187-8795, USA.

Steve Kroll-Smith is Professor of the Department of Sociology and Director of the Environmental Social Science Research Institute at the University of New Orleans, New Orleans, Louisiana 70148, USA.

Anthony Oliver-Smith is Professor of the Department of Anthropology and Co-Director of the Displacement and Resettlement Studies Program, 1350 Turlington Hall at the University of Florida, Gainesville, Florida 32611-2036, USA.

Ronald W. Perry is Professor of the School of Public Affairs at Arizona State University, Tempe, Arizona 85287, USA.

Boris N. Porfiriev is a Leading Research Fellow and Professor of the Institute for System Studies (VNIISI), Prosp. 60 let Oktiabria 9, 117 312 Moscow, Russia.

E. L. Quarantelli is Research Professor of the Disaster Research Center at the University of Delaware, Newark, Delaware 19716, USA.

Uriel Rosenthal is Professor of the Department of Public Administration in the Faculty of Social Science and Law, PO Box 9555, Leiden University, 2300RB Leiden, The Netherlands.

Robert A. Stallings is Professor of Public Policy and Sociology at the School of Public Administration, University of Southern California, Los Angeles, California 90089, USA.

FOREWORD

In 1917, a Canadian named Samuel Henry Prince began the formal study of Sociology of Disaster with his dissertation on Canada's worst catastrophe, the 1917 Halifax explosion. However, it wasn't until after the Second World War that scholars began to concentrate on this area and produce perceptive studies such as Charles Fritz and J.H. Mathewson's *Convergence Behavior in Disasters* (1957), readers such as George Baker and Dwight Chapman's *Man and Society in Disaster* (1962), and books such as Allen Barton's *Communities in Disaster* (1969) and Russell Dynes' *Organized Behavior in Disasters* (1970).

In the same era, Dynes' long-time colleague, Henry Quarantelli, started pressing for a world-wide organization of disaster scholars and insisting that this body must have a journal. Quarantelli's efforts led to the creation of the Research Committee on Disasters, International Sociological Association, and publication of the *International Journal of Mass Emergencies and Disasters*.

As those who come to know him well learn, Quarantelli is not the sort of person who is ever satisfied with what has been accomplished. He is always searching for new subjects to study, new ways of sharing what has been learned. This volume is a direct result of his dedication.

I want to thank Henry Quarantelli, not just for his work on this book but for the decades of devotion to his students, his colleagues, and, generally, to the world disaster community. There are very few persons who can honestly say that, within their lifetime, they have watched a field develop and prosper and that they have played a key role in making that possible. Henry Quarantelli can say that. All of us in this field owe him our deep gratitude.

Joseph Scanlon
Director, Emergency Communications Research Unit
Carleton University, Ottawa, Canada

President, Research Committee on Disasters

ACKNOWLEDGMENTS

We want to express our appreciation to all the authors, not only for their original and fine contributions, but also for their willingness to waiver any possible royalties to the ISA Research Committee on Disasters. In turn, the Committee is thanked for obtaining an agreement with Routledge to develop a series of books on the human and social aspects of natural and technological disasters, something lacking up to now except for efforts in some academic circles that of necessity were limited in the audiences that they could reach.

In addition, we want to thank all the authors whose native language is not English. As is well known, subtleties are often lost in translation. This has undoubtedly happened in some of the articles since all authors wrote their final drafts in English. It should also be noted that in editing these manuscripts for publication, we did leave in place as much as possible even some awkward constructions and phrasings, so as to keep as close as possible to the original versions written by the authors.

ELQ

1

INTRODUCTION

The basic question, its importance, and how it is addressed in this volume

E. L. Quarantelli

What is a disaster? That is the question that this volume addresses. A dozen answers are posed by researchers from six different social science disciplines and from half a dozen different societies.

The background of this particular quest is as follows. From my earliest involvement with research in the area initiated more than four decades ago, I have struggled with how to define and conceptualize the term "disaster." This concern has provoked me at different times through the years to advance various conceptions and to analyze what others were proposing (as examples, see Quarantelli 1966, 1977, 1985b, 1987a, 1987b, 1989a, 1989b, 1992a, 1993a, 1993b, 1994). Our starting point was the varying ideas set forth in the earliest days of systematic disaster research in the social sciences, in the 1950s and 1960s. Among the more relevant ideas were those expressed by Endleman (1952), Powell, Rayner and Finesinger (1952), Killian (1954), Williams (1954), Moore (1956), Fritz (1961), Barton (1963, which preceded his later better known work, 1969), and Stoddard (1968). There were also some earlier ideas expressed by Carr (1932) and Sorokin (1942), while known to a few, were very seldom explicitly acknowledged by most of the pioneers in disaster studies.

At one point I even traced historically some of the earlier discussions and formulations of the term. In particular, I noted a move from the use of a label with a referent to primarily a physical agent to one which mostly emphasized social features of the occasion (see Quarantelli 1982). However, while the indicated writings as a whole represented initial worthwhile clarifications of the problem and a movement in the right direction, they still fell far short in my view of what was needed, especially for social science research purposes. In fact, as I evaluated the field about five years ago, it did not appear to me that the overall situation had materially changed that much since my earlier analysis about a decade earlier.

However, in 1992, the International Institute of Sociology requested papers for its Congress to be held at the Sorbonne in Paris in June 1993. I seized upon this opportunity to propose a session on the concept of "disaster." Since European scholars had seemed consistently and proportionately more interested in the problem than their American counterparts (e.g. Westgate and O'Keefe 1976; Clausen, Conlon, Jager and Metreveli 1978; Ball 1979; Dombrowsky 1981; Pelanda 1982a, 1982b; Schorr 1987), I asked a number of them to prepare papers for the meeting. In very general terms, they were asked to put together a statement on how they thought the term "disaster" should be conceptualized for *social science research purposes*, and to indicate at the same time their questioning and/or criticizing of past and existing conceptions. The session was entitled: "Disasters: Different Social Constructs of the Concept."

Some of those invited could not participate, but eventually I selected five papers, four by Europeans and one by an American. At the meeting itself, four papers were presented: by the sociologist, Dombrowsky, from Germany; by the political scientist, Gilbert, from France; by Horlick-Jones from Great Britain, at that time in the Department of Geography at the London School of Economics; and by Gary Kreps, a sociologist from the United States. I particularly asked Kreps because he has been one of the few Americans who, throughout his professional career, has explicitly expressed an interest in the question of how to define and conceptualize "disaster" (e.g. Kreps 1978, 1983, 1984). Porfiriev from Russia, with a doctoral degree in economics, was unable to come to the meeting as he had intended, but later, after reading the other four papers, also wrote a manuscript.

As coordinator of the roundtable, I did not see it as a proper role for me to present my own point of view at that time on the definitional/conceptual problem (although I did so later in a paper presented at the World Congress of Sociology in 1994 (see Quarantelli 1995b)). However, to understand the background for the papers that were presented, it does seem appropriate for me here to restate in summary form my opening remarks to the roundtable. This is what I said.

Why did my proposal about the roundtable come when it did? Because in my view it is time after nearly half a century of fairly extensive empirical disaster research, to systematically address the central concept of the field. In the early stages of the development of any scientific area it matters little what researchers do and explore. Almost anything empirically found is worthwhile discovering. However, studies in the sociology of knowledge suggest that after a certain period of pioneering work, a developing field will flounder unless there emerges some rough consensus about its central concept(s). Thus, my view is that unless the field of disaster research comes to more agreement about what a disaster is, the area will intellectually stagnate (in another recent paper, I expound more on this thesis, see Quarantelli 1995b).

Empirical work of course could and would continue, but without the intellectual infrastructure and scholarly apparatus any field needs (by way of explicit models, theories, hypotheses, etc., all of which however require addressing conceptual issues), the research will continue at best to produce only low-level empirical observations and findings, without any significant accumulation of systematic knowledge. (The research funding in most societies unfortunately tends to support very strongly only such low-level applied work rather than more abstract basic research.)

The term "disaster" is certainly the key concept in the area. Yet even what is assumed in the subtitle of the roundtable—namely, *different social constructs of the concept*, is not fully agreed upon or used. Many in the field assume there are physical happenings out there, independent of human action in any sense (e.g. most, although not all, geographers assume that to have a disaster there must be the physical presence of a hazard, i.e. an earthquake, flood, cyclone, etc.). If workers in the area do not even agree on whether a "disaster" is fundamentally a social construction or a physical happening, clearly the field has intellectual problems. (I might note that even formulations by sociologists and other social scientists that appear to be fully social constructs, but which use geographical space and/or chronological time as dimensions or factors in defining a "disaster," in my view, are not using fully social construction concepts. As I will discuss in the final chapter, they should use social space and social time features, ideas well developed by some sociologists since Sorokin and Merton long ago urged their use.) So a major reason we need clarification is because otherwise scholars who think they are communicating with one another are really talking of somewhat different phenomena. A minimum rough consensus on the central referent of the term "disaster" is necessary.

In saying this, I should stress that in no way am I arguing for agreement on one single, all purpose, view of disaster. As noted even in some of my writings of several decades ago, for legal, operational, and different organizational purposes, there is a need for and there necessarily will always continue to be different definitions/conceptions of when a "disaster" is present. I have no problem with such different views about what constitutes a "disaster."

However, in my view, for *research* purposes aimed at developing a theoretical superstructure for the field, we need greater clarity and relative consensus. For example, some disputes about what appear to be empirical findings mostly stem from different usages of the basic concept in the field. For instance, whether there are serious negative mental health consequences of X (disasters) is hotly disputed, because some of the researchers use such a broad referent that any type of individual or group stress situation is seen as a "disaster" (see Quarantelli 1985a). Of course, the broader the referent the more likely it will be that one will find any given phenomena. Some of us who use a narrower referent for "disasters" exclude a wide range of stress situations, such as conflict situations (e.g. war, imprisonment in concentration or military camps, terrorist attacks, riots and civil disturbances, hostage takings, etc.), where we would not disagree there could be severe negative mental health consequences. Thus, the dispute in this case is not mostly about the empirical findings, but the referent of the basic term "disaster." The same can be said about antisocial behavior. If conflict situations are treated as "disasters," by definition one will have antisocial behavior. That conclusion results not from empirical findings, at least at one level, but from the definitional referent of the basic term "disaster."

So to be concerned about what is meant by the term "disaster" is not to engage in some useless or pointless academic exercise. It is instead to focus in a fundamental way on what should be considered important and significant in what we find to be the characteristics of the phenomena, the conditions that lead to them, and the consequences that result. In short, unless we clarify and obtain minimum consensus on the defining features per se, we will continue to talk past one another on the characteristics, conditions and consequences of disasters. In concrete historical terms, would all of us agree or not agree that what is currently happening in Somalia and in Bosnia should be treated as "disasters?" From previous discussion among us, we know that we do not fully agree on the answer to that question. There is not agreement because we have different ideas of what should be the defining features of the basic concept of the field.

Finally, as a last point, I would mention that we require clarification because we are also at the threshold of the appearance of certain kinds of relatively new social happenings that will need to be either included or excluded from the rubric of "disaster." Examples are such phenomena as the AIDS epidemic, computer and high tech network failures, biogenetic engineering mishaps, and accidental as well as deliberate large-scale food and drink poisonings and contaminations. What is noticeable about these and similar happenings are that they usually occur independent of particular local communities, can be characterized in terms of social space and time, and often do not result in any or many sudden fatalities/casualties or significant property damage. Yet, they create major social disruptions, are economically costly, and can be very psychologically disturbing. Or, to use a more familiar example, was the Three Mile Island nuclear plant accident a "disaster" in the same sense as was Chernobyl?

Some see the former as not a disaster at all, given the absence of casualties from the occasion, while others such as myself see it as a harbinger of future "disasters." Then there are also a myriad number of happenings frequently captured under the label of "industrial crises" which only partly overlap at best what more traditional "disaster" researchers study (see Mitroff and Kilmann 1984; Shrivastava, Mitroff, Miller and Miglani 1988). Which, if any, of these happenings should also be in the bailiwick of disaster researchers?

The above are paraphrased comments on what I said in opening the roundtable at the Sorbonne. There was no anticipation on my part that the ensuing discussion would result in consensus even among the four researchers involved. This expectation proved correct.

However, the papers and the discussion at the session led me to think it might be worthwhile to try not only to continue but to expand a dialogue among those scholars interested in the topic. This was reinforced by an encounter at a professional meeting in Mexico with Ken Hewitt, a Canadian geographer. In 1983 he had written one of the most detailed and explicit criticisms of the field of disaster studies up to that time. He raised, in my view, very trenchant questions about the definitions and conceptions of disasters that had been and were being used in the social science literature. I proposed to Hewitt that he write a reaction paper to the other five papers.

The outcome of all this was presented in a special issue (that of November 1995) of the *International Journal of Mass Emergencies and Disasters* (IJMED). The issue consisted of the original five papers (with the ones by Horlick-Jones and Kreps being updated since their oral presentation in Paris), and the reaction paper by Hewitt. In addition, I asked each of the original five authors to write a short reaction paper to what Hewitt had said about their own individual contributions. The format used of having original papers, a general reaction paper, and reactions paper to the general reaction paper was my effort to generate something of a dialogue between the scholars involved.

As might be expected, some authors in their papers were circumscribed in explicitly criticizing in any way the views of others. However, fulfilling my hope, some did specifically, directly, and candidly address what others had written. The purpose, of course, was not to generate conflict but to force the writers to more explicitly expand on their own positions.

Even at that time I was aware that there were important and different points of view regarding the conceptualization problem that were not clearly represented in the special issue. Dynes, for example, had advanced a new and novel perspective on the whole problem (see 1994a, 1994b). In addition, there were others, such as Rosenthal and his colleagues (1989a, 1989b), Oliver-Smith (1993) and Mitchell (1990, 1993), who had written statements that were only partly reflected in the journal issue.

This led me to the idea to have a second round of papers that would follow the same format as the first round that appeared in the journal issue. I asked five other prominent disaster researchers (Dynes, Kroll-Smith who added a co-author in Gunter, Oliver-Smith, Rosenthal, Stallings) also to write original essays, and to include their views about the writings in the first round. In turn, doing the same as Hewitt had done, Ronald Perry, who had accepted my invitation to do so, wrote a reaction paper on the five new papers. My choice of Perry was strongly influenced not only by the excellence of his research in the disaster area, but also by my awareness of his encyclopedic knowledge of many theoretical and methodological issues in sociology and related social science disciplines. Again, paralleling what had been done before in the first round, each of the authors also wrote a brief reaction note to Perry's article. All of these appear in this volume. The exception is the article by Horlick-Jones that I dropped from the collection.

In addition, there are (revised) introductory and (original) epilogue chapters by myself. All the separate references in each of the papers have been consolidated into one master bibliography included at the end of the volume.

My intent was to obtain as many different disciplinary perspectives from the social sciences as possible. I was successful in obtaining contributions from scholars whose backgrounds were in anthropology, economics, geography, political science and public administration, and social psychology and sociology. Psychology is the only major field not represented, but then explicit and extended theoretical discussions of what is a disaster—although not absent—are not common in that discipline (most theoretical discussions are only peripheral in connection with larger discussions of disaster-related mental health problems; for example, see Berren, Beigel and Ghertner 1980; Wright, Ursano, Bartone and Ingraham 1990). Now, whether the provided contributions truly and fully represent the indicated disciplines is a question others might want to consider. However, I would argue that at least part of the intellectual dialogue that took place actually reflects different views both within and between disciplines on how disasters should be conceptualized.

To a degree, the authors are from those countries where the greatest amount of social science studies of disasters are currently being undertaken. The major exception is Japan and a few developing societies where studies have accelerated in recent years. Unfortunately, my efforts to obtain contributions from these places did not work out. However, to maintain proper perspective it should be noted that while there is much Japanese empirical research, theoretical work is not prominent in that country, and the great majority of studies in developing societies are of a very applied nature. Nevertheless, in my concluding chapter, in which I make some suggestions as to where the field of disaster research ought to go in the future, I do indicate why work from developing countries might soon be a very good starting point for suggesting a radically different or revolutionary view on the question of what is a disaster.

My choice of different disciplinary backgrounds and national scientific circles was an attempt to mirror in some crude way the current research setting in the disaster area. To the extent that my rough sampling was anywhere near valid, the papers written for this volume provide a gross reflection of how social science researchers think about disasters. Or, perhaps more accurately, the articles in the rest of this volume reflect the perspectives of those in the field who have most explicitly and consciously thought about the central concept of the field.

Finally, it should be noted that the first set of authors who wrote for the IJMED journal were asked to react primarily to the basic question: What is a disaster? On the other hand, the second set of authors were asked to indicate their positions not only about the basic concept, but also additionally what they thought about what the original five authors had written. As such, the last set of authors wrote whatever they did within a larger reactive framework, and therefore as a whole tended somewhat to discuss broader issues in their articles than those who had only originally written their papers for the IJMED. Thus, such differences as are manifested in the number of matters addressed are more a function of the two formats within which each set of authors had to write.

Part I

QUESTIONS IN THE STUDY OF DISASTERS

There are nine papers in this part. For the most part the authors here discuss either past or current views about the concept of disaster. Using a historical approach, Gilbert argues that the central concept has undergone at least three major changes since disasters have become the focus of study. Dombrowsky discusses why it appears that what is a disaster is a question that is asked again and again. In a very systematic presentation, Kreps sets forth his particular view of disaster as a systemic event and social catalyst. Porfiriev notes that the concept of disaster is rooted in two approaches to research, namely an applied/pragmatic one and a theoretical/conceptual one.

Hewitt then presents his reaction to the four papers and goes on to discuss what he considers current perspectives that are excluded from these formulations. Each of the four original authors then present their own brief reactions to the Hewitt paper.

2

STUDYING DISASTER

Changes in the main conceptual tools

Claude Gilbert

The numerous theoretical approaches to disasters can be classified into three main paradigms. We present their content, chronological developments and cleavages. The first is disaster as a duplication of war (catastrophe can be imputed to an external agent; human communities are entities that react globally against an aggression). The second is disaster as an expression of social vulnerabilities (disaster is the result of underlying community logic, of an inward and social process). The third is disaster as an entrance into a state of uncertainty (disaster is tightly tied into the impossibility of defining real or supposed dangers, especially after the upsetting of the mental frameworks we use to know and understand reality).

In his review of the main theoretical approaches to the field of disaster, Quarantelli (1985b) in an ironic manner questioned scientists about *what a disaster is*. He also pointed out that the larger debate among scholars, as well as the mainstream scientific works in the field, followed a common "fault line" which has long been a major weakness in its theoretical foundations. While Quarantelli's challenge has since given considerable impetus to scientific exploration, it has also resulted in restraining the profusion of works carried out on a case study basis. The field of disaster studies has long been overly dominated by narrow analysis of particular cases, and it has therefore been lacking in theoretical investigations on foundations and concepts.

This paper represents an initial attempt to elucidate the main conceptual changes or even cleavages that have been taking place in disaster analysis in recent years, as well as those which are likely to emerge from the present debate. Our discussion will be no more than a skeletal research framework that is in need of further development. But a critical synthesis will be possible only if one accepts a summation, maybe in a simplistic way, of the minutiae and the controversies of a larger scientific debate. This paper looks briefly at the conceptual scope of the field as a whole, and it follows the chronological development of the principal explanatory approaches on which disaster analysis has been based. The discussion presents the case for reordering the theoretical foundations in the study of disaster, thus enhancing a real conceptual debate.

PARADIGM 1: PATTERNS OF WAR APPROACH

War has long been the subject of exploration by social scientists and social analyses have easily found war patterns. Such patterns are particularly relevant to the field of disaster studies. The patterns have long been viewed as being the result of harmful attacks brought against human groups.

In this sense at least, disasters bear a great resemblance to war, with the causes of disasters being sought outwardly. With the concept of *agent* being used to refer to both arms and enemies, disaster has since the beginning been explained on external grounds. As a result, human communities have been seen as organized bodies that have to react organically against aggression. Actually, the external agent was found to find the group, as Hegel said of God and the world. Another outcome is that the agent and the community are linked and that this linkage is one of conflict, the agent being the source of the action launched against the community.

This paradigm still holds true, mainly because it is simple and clear. The fact that war and disaster have something in common is clear, and this evidence is part of our common sense. Researchers themselves can hardly approach disaster through explanatory tools different from war patterns, which have long since been the standard criteria for talking about disaster, and making it immediately intelligible to everybody. As our discussion will later show, political and public servants, as well as the representatives of citizens and scientists tend to be trapped by the traditional overemphasis on a war rhetoric, through which a disaster, however complex it might be, can take material form.

If this paradigm still persists in a field of research enriched by criticism, like other fields of research, it is neither only due to the resemblance between disaster and war, as it is mentally represented, nor to the effective discussions of war.

The paradigm of war patterns strongly reflects the circumstances and the place where it first emerged. It was in the United States at the height of the cold war. Fritz (1968) has argued that US government institutions provided research funds at that time primarily for studies relevant to understanding the reactions of people to possible air raids. Disasters were viewed as situations likely to elicit the reactions of human beings to aggressions and to allow an adequate test of them. The scientific approach to disaster is therefore a reflection of the nature of the *market* in which disaster research became an institutional demand. Bombs fitted easily with the notion of an *external agent*, while people harmed by floods, hurricanes, or earthquakes bore an extraordinary resemblance to victims of air raids. This was particularly true in the United States where natural disasters, as compared to technological and industrial catastrophes, took place frequently. It thus can be argued that the study of disaster at its beginnings was strongly related to the nature of the institutional demand for it.

The points mentioned above hold true for disaster studies in other countries also. In France, for example, institutional demands for disaster relevant research (i.e. the crisis sector) comes mainly from the successors of civil defense or civil security agencies created during the two world wars. Even though these agencies now are concerned with what in France is called "Security," they emphatically remain consistent with military patterns. As a result, they naturally tend to interpret disasters through war patterns while shaping theoretical analyses and expertise studies through the approach traditionally overly dominated by war.

To sum up, the success of this paradigm can be explained by the nature of the demand that helped disaster studies emerge, and by formal as well as analytical factors. However, recently the war pattern approach has been challenged, and it has been the subject of a larger critical debate among American academic commentators. A major cleavage in the field was brought about by Quarantelli (1970b) and his concept of *consensus crisis*. What matters in the conceptual shift introduced by Quarantelli is, first, the change he carried out within the common interpretations of disaster and, second, the explicit criticism he made of the unnecessary linkage between the destructive factor and the community as it emerged from the notion of panic. He pointed out that there was no mechanical relation between these two factors, and thus that there was great autonomy in the reactions of people to troubles. At the conceptual level, Quarantelli triggered the emergence of new modes of approaching disasters, based on an analysis of communities, and not only of destructive external agents.

Quarantelli largely paved the way for a new type of research in the field of disaster studies. Wenger (1978), for example, has recognized that social factors within communities were also relevant to the understanding of disasters. As a partial result of this shift in the conceptual approach to disaster, destructive agents are no longer seen as a cause, but as a *precipitant for crisis and disaster behavior* directly related to the social context. At the same time, Dynes (1974) has pointed to the social fracture triggered by disasters, while Kreps introduced the concept of *twin disaster*. While approaching disaster in relation to external destructive agents, US researchers made an important turning point in the 1970s in conceptualizing disasters and left the mainstream crisis paradigm. They were forced, in varying degrees, to recognize that disaster had to be studied *within* the human group involved in it, and not as the result of an exclusive external factor. Though deserving scientific debate and refreshing in its breadth, US research was unable to enhance a completely critical analysis. This was in fact carried out by European scientists.

PARADIGM 2: DISASTER AS SOCIAL VULNERABILITY

The second paradigm was critical of the failure of the first approach to utilize the insights generated by further scientific exploration in the field of disaster studies. The strengths and weaknesses of the war pattern approach have been assessed in detail.

Some authors are worth mentioning. Jaeger (reported in Schorr 1987) studied the case of a community pushed together by an external danger. He demonstrated that this type of group followed a community or family reshaping that is common in American society. Dombrowsky (1981) clearly ascertained the extent to which disaster analysis was dependent upon war patterns, particularly those of thermonuclear war. At the same time, however, he shows that the underlying logic of disaster studies was intrinsically seen as related to the external destructive agent, and as a partial result, that the reactions of people were the only indicator of the nature of the agent. He suggests a creative reformulation in studying disaster as a social action taking place within societies. At a more radical level, Pelanda (1981) echoes Dombrowsky, in that he considers disaster as a social result and a consequence of sociostructural risks. And he poses the question as to whether the notion of disaster itself is to be totally rejected, in order to assess the real importance of uncontrolled processes on social vulnerability.

The insights generated by this criticism in the late 1970s and the early 1980s triggered new modes of approaching disaster, which have increasingly replaced the traditional paradigm. The new approach to disaster not only reverses the old hierarchy of factors, but gets rid of the overwhelming notion of *agent*. Starting from an analysis of disaster seen as a process tightly tied to social vulnerability, the new paradigm considers that the causes of disaster are to be explained on structural as well as contextual grounds. When social risks explode that are totally raised inward and not outward into the community, then there is a disaster.

As a result of this first conceptual shift, disaster is no longer experienced as a reaction; it can be seen as an action, a result, and, more precisely, as a social consequence. The new approach provides the basis for moving from disaster as an effect to disaster as a result of the underlying logic of the community. Another outcome of the change in paradigm implies that with social risks exploding in a community, disaster is virtually experienced as a process whereby specific area activities carried out by the actors and the structures of the community begin melting. Therefore, the conceptual framework of disaster is neither one of conflict, nor of defense against external attacks, but is the result of the upsetting of human relations.

The new approach, as it has since been developing in contrast to the old paradigm, is based on theoretical foundations similar to those espoused in 1932 by Carr (1932). However, while the raising of a larger scientific debate was good, the introduction of a new paradigm has raised a large range of new questions. As yet, criticism of the old paradigm, though accurate, lacks any unifying conceptual focus able to make new suggestions.

The first obstacle to the new paradigm is that it gets rid of the notion of "agent," thus breaking a comfortable standpoint with common sense. As we put it at a meeting in Grenoble, it is simpler to say that a shipwreck was caused by a storm, than to explain that the risks of a ship and its crew were revealed during a storm (Gilbert 1991). It is not easy to introduce change into common sense, particularly when this change triggers a cleavage whereby actors and institutional agencies do not share the new mode of approaching disaster and are therefore cut off from scientists. At the same time, researchers are not unanimous in support of the new paradigm, and some of them keep on considering dysfunction or accident as central to disaster. Among US scholars, this position seems to be common. Lagadec (1988) clearly separates accident from crisis, even though he argues that accident actually is the first and most important factor triggering what we call "post-accidental" crises. Behind these attempts to solve the question of approaching disaster, there appears to be a theoretical problem for relating the two paradigms and shaping a unifying research framework wherein anyone might find a scientific basis for his work.

The second obstacle is related to the notion of vulnerability. What is the meaning of *vulnerability*? Is it related to social organization and, therefore, does it rely on the nature of compromise or consensus between social groups? If so, the approach assumes, as do the works of Quarantelli, that social disorder and panic still remain central to disaster analysis. If vulnerability has political implications and responsibilities, it directly concerns the ability of political actors, sometimes called decision-making actors, to face critical situations. In both cases, the conceptual direction is traditional in that the question still is whether actors are able to make decisions. But there is something new in this approach: the reason for the production of disorder is to be found within society and not from outward forces. But, for both social and political vulnerability, the background is the same: the crash of a community is explained by the relaxation of social and political boundaries.

Truly, the authors of the new approach have sometimes left unsolved the question of vulnerability. For Pelanda (1981), however, who follows Perrow, vulnerability is a key issue in a set of uncontrolled processes. With the uncontrolled processes becoming a major explanatory tool for understanding vulnerability in complex societies, danger has increasingly been linked to organizational modes in modern societies, and particularly in those sectors where activities are dangerous or fatal. But the criteria for gauging the extent of risk in complex societies are still traditional: human losses, damages, and the risks of upsetting social and political order, remain the main standard parameters.

Actually, the theoretical scope of the new paradigm has directly been affected by the old approach, which is just its opposite. Arguably, there are overt signs that a third paradigm is shaping new contours in the field of disaster studies, cutting across old distinctions and traditional concepts and paving the way for a different definition of "disaster."

PARADIGM 3: DISASTER AS UNCERTAINTY

What has led to a significant shift from old approaches to a renewed vision of disaster, is the tendency to consider disasters as crises (see Lagadec, as well as Rosenthal and others). The first steps were uncertain, with the external factor still being used as an explanatory tool, and the autonomy of crises being ascertained through a clear distinction between accident and crisis. At this stage, the suggestion was made that accidents could occur without any crisis, and that crises could emerge without any accident. However it may be, disaster is clearly studied through the crisis that develops within a community. Although still imprecise, this notion helped reinterpret disaster as a serious disorder taking place within communities and, most of all, as a disorder triggered by communication problems (see Lagadec 1988). Hence, disaster is first of all seen as a crisis in communicating within a community—that is, as a difficulty for someone to get informed and to inform other people (Gilbert 1991).

Although refreshing on the whole, in that they utilize the notions of crisis and communication, these approaches are not completely new. Fritz had already pointed to a third explanatory factor for disasters, what he called the upsetting of *the system of meaning*, suggesting that communication influentials are overwhelmingly relevant to the interpretation of confused or chaotic situations. However, at the theoretical level, a real turning point was reached when disaster was first related to the beginning of uncertainty. Authors such as Dynes and Quarantelli had previously mentioned uncertainty, but only recently has Rosenthal's work (1989) overtly and fully expressed the idea that uncertainty, as tightly linked to menace, is at the core of disaster theory and an important indicator of the range of crisis factors.

Uncertainty has long been the subject of social science exploration, as in studies on organizations, public policies, or international crises. It is worth posing the questions, however, as to why the notion of uncertainty has emerged so late in the field of disaster studies, and also whether this delay has affected the explanatory potential of uncertainty in the system of meaning. A powerful reason as to why the notion of uncertainty is needed to understand the development of technological accidents and natural catastrophes is mentioned in recent studies carried out in France (Gilbert 1991).

Uncertainty has been indicated as a production of complex societies, and not as a result. Following this approach, the process of uncertainty is engaged when there is a disappearance of the cleavage between the scopes of activities in which actors and structures are used to operating (see Dombrowsky 1985). Uncertainty virtually occurs when actors and structures stop developing their own underlying logics and begin operating only on the border of their respective areas of activities. At the same time, they freely take part in the crisis market acting in disorder, and try to define the situation. One outcome of the convergence of specific areas of activities in which the actors and structures usually operate is a change in ordinary communication networks and knowledge standpoints, with competence areas and mental habits being totally upset.

Uncertainty is not merely due to the absence of communication or information in modern societies. It is the anarchical profusion of information that is responsible for uncertainty, since it deeply affects the system of meaning that is tightly linked to the modes of organization of administrative, political, and scientific fields. As a philosopher put it: disaster is less an accident of reality than a disaster is the representation of reality.

Starting from the points mentioned above, the third paradigm can be described in terms of three points. First, disaster is tightly linked to uncertainty that occurs when a danger, whether real or not, threatens a community, and this danger cannot be defined through causes or effects. Second, uncertainty emerging from modern communities is related to their growing complexity. It is the result of the upsetting in the system of meaning, and not the effect of the difficulty of solving problems of accidents or serious dysfunctions. Uncertainty is mainly the product of community organization and not of external factors. Third, we may speak of disaster when actors in modern societies increasingly lose their capacity to define a situation that they see as serious or even worrying through traditional understandings and symbolic parameters. For a community, disaster means the loss of key standpoints in common sense, and the difficulty of understanding reality through ordinary mental frameworks.

Criticisms directed at the new paradigm likely to bring about deep changes in the approach to disaster are already known. An approach based overwhelmingly on the system of meaning seems likely to underestimate other factors such as human losses, damages, social and political disorders, and the moral trouble which still constitute the reality of disaster for many researchers. Basically, it is worth asking whether it is true that human communities end up accepting anything, when they are able to interpret facts through different conceptual tools, however technical, scientific, symbolic or even magical they might be. And, reversing the terms, whether or not something cannot be accepted when people cannot be informed about a danger. The tragic experience of what happened in Hiroshima indicates that the first suggestion might have some validity. But this question has long been the subject of exploration, and not only in the social sciences.

CONCLUSION

This paper has argued that a distinction has to be made between the three approaches to disasters. The discussion on paradigms needs further development, particularly to show that they have provided bases for working out conceptual patterns that had strong reliance in each of the three approaches. It would also be worth ascertaining, as Quarantelli has done, the extent to which critical analysis in this field could productively have developed, and asking why the insights generated in other similar fields have been rarely, or only slowly, taken into account in disaster studies. It has been pointed out that contemporary societies are growing in complexity (as, for example, the sociologists of organizations and the analysis of complex systems have done). It has been shown that the impact of cognition processes is also relevant in the field of disaster studies (for example, in psychosociology first, and later in the cognition sciences, given the strong development they have both experienced), as well as the impact of symbolic representations (i.e. anthropology) and the position accorded to power by political scientists within modern democratic societies.

The field of disaster studies has rarely taken into account contemporary theoretical debate. One of the possible reasons that may be suggested is that disaster is one of those ultimate things that can be thought of only within a larger framework involving the city as a whole. The meaning of disaster is to be found within the questions directly concerning the organization of human communities: what kind of links of solidarity can make a community out of different social groups? What is the meaning of the responsibility of the public authorities in economic, political, administrative, technical, or scientific areas? What is the convergence between the general and the private interests, between the long and the short terms? What is the right form of governing ordinary and exceptional situations, the correct compromise between the "raison d'etat" and democracy? Questions concerning disaster and the conditions in which they occur, directly affect the foundations of human societies.

Such subjects have the peculiar characteristic of raising questions that cut across scientific debate. To put it in Quarantelli's words, the fact that the definition of disaster has been so slow to emerge can be explained by the different utilizations of the concept of disaster out of the scientific field. The different meanings of the notion of disaster have a strange impact on the concept itself. This clearly makes us understand why the first paradigm, though not scientifically tested, has resisted so long. As a partial result, the third paradigm, which is critical of common evidence, has not as yet been able to dominate the realm of disaster studies.

3

AGAIN AND AGAIN

Is a disaster what we call a "disaster"?

Wolf R. Dombrowsky

Following Carr, who defined disaster as the collapse of cultural protections, this paper develops a sociological approach to processes commonly called a "disaster." Epistemologically, the definitions used in science and practice are classified and redefined as *programmatic declarations*. Definers declare what they perceive as a problem and how they intend to solve it. Given the fact that neither "problem and perception" nor "solution and exigency" necessarily match, the probability of mismatches increases when inconsistent conceptions prestructure the view one has of reality. Still, the transformation of nature into culture is interpreted within a premodern expression and false causal attractions: "Des Astro," "evil star," "bad luck" and "blind faith." In contrast, this paper suggests a conception that defines disaster as an empirical falsification of human action, as a proof of the correctness of human insight into both nature and culture.

Similar to the beginning of sociology as a science, the subdiscipline of disaster sociology faces the problem of defining its object of study. To myself, already, the distinction between two different classes of objects, *natural* and *manmade disaster*, seems fairly unsociological. Moreover, the definition of disaster as an event raises more questions than any sociological elucidation. In contrast to other scholars in the field, I suspect more dissent than consensus in the ways of conceptualizing the domain of our object of study. The vast number of definitions of "disaster" as an event (Fritz 1961: 655) or as an acting entity (Kreps 1993: 6) may be mistaken as consensus. But it should be seen as the outcome of a scientific tradition that is "concentrated in time and space." It is an American specialty, developed and elaborated during the postwar development of the sociology of disaster (see Anderson 1979). However, an emancipation of the field from everyday knowledge and from the practical needs of disaster management has been neglected during this phase of its establishment. Up to today, there is no epistemology of the sociology of disaster. Consequently, almost no sociological definition of disaster does exist. Thus, instead of harmonizing the views in the field, I will explicate as pointedly as possible my understanding of a sociological definition of disaster, because from a European perspective there still is a lack of sociology in sociological disaster research (Pelanda 1982a; Gilbert 1992).

WE SEE WHAT WE WANT TO SEE

In his epistemological attempt to conceptualize disaster, Quarantelli stated: "We all have a habitual way of looking at disaster phenomena" (1982: 453). So have I and so have all others. Westgate and O'Keefe (1976) analyzed circulating definitions and found that they were mere programmatic declarations. Those who define declare what they intend to do with the social processes called disaster. This is different from defining disaster. The German Red Cross, for example, defines disaster as an "extraordinary situation in which the everyday lives of people are suddenly interrupted and thus protection, nutrition, clothing, housing, medical and social aid or other vital necessities are requested" (Katastrophen-Vorschrift 1988: 2). The German law that states the laws of disaster protection itemizes phenomena (such as storm, flood, blizzard, explosion, etc.) which are seen as typical in releasing disasters. However, it defines "disaster" almost in the same tenor by saying it involves "such severe interference of the public order and safety that an intervention of the centralized, coordinated disaster protection units is necessary" (Seeck 1980:1). German insurance companies define disaster as a situation involving damage and/or loss of lives beyond one million German marks and/or 1,000 persons killed.

It is easy to add more examples of definitions that serve for nothing else than to claim that the definers approach reality under specific conditions. For the Red Cross, a disaster is a large-scale lack of nutrition, clothing, housing, aid, etc., or roughly summarized, a situation where the services and offers of the organization are heavily demanded. Even more tautological is the definition of disaster by law: A disaster is what the intervention of disaster relief units makes necessary, and due to legal construction, what was called forth by coherent "triggers." For the state, the breakdown of public order and safety is the key, not the phenomena itemized. However, the specification of possible disasters is required because of the need for an appropriate selection of countermeasures to reestablish public order and safety. If public safety is threatened by other triggers, other—yet appropriate—countermeasures have to be selected. The trigger determines the measure; thus, riots, the use of the National Guard; epidemics, the General Surgeon; terrorism, the Special Forces and Bomb Squads. The maintenance of public order and safety has to be guaranteed under all circumstances, no matter what phenomenon has caused the trouble. The type of phenomenon is only the key for the use of the appropriate toolbox.

For other definers, the function of definitions is the same. For the Red Cross, disasters are large-scale situations that lack vital necessities and triggered by sudden interruptions of the everyday lives of people. But in contrast to the legal context that uses a trigger attributable to a phenomenon known as disaster, and a specific activity to maintain the public order, the Red Cross does not even need a precise differentiation between trigger and activity. On the contrary, sometimes the wrong trigger produces a lack of vital necessities, which is the case when charity is hampered by repulsive pictures (of victims who have drowned in floods, for example) instead of being spurred by pitiful pictures (such as children victimized by droughts and starvation).

TYPES OF DEFINITION

With all that in mind, a strange conclusion may be drawn: the term "disaster" has only ephemeral significance. It is a trigger, a flag to signal a meaning, a stimulus to produce a specific reaction. Yet it has almost no importance for the activities that are carried out under the label of a disastrous event. Right here, different lines of argumentation have to be followed up. The first line is following the question of how language is structuring our perception of the world; the second line is following the question of how reality is transformed into the mechanics of problem-solving; and the third line is following the question of how disaster sociologists and their ways of conceptualizing disaster will be affected by all this.

Let us trace the first line of argument. I wonder at my own easiness in using the term *event* in the paragraph above. (Mis)using this term, the nesting of the concept of disaster as an event in our minds becomes as obvious as the difficulty of avoiding such a view. It is much easier to freeze a complex social process into a static actor or a "thing," than to express adequately its dynamic complexity. However, a conception of casualty that describes nonhuman occurrences in terms of human activities is usually called animistic. Phrases such as "a disaster hit the city," "tornadoes kill and destroy," or a "catastrophe is known by its works" are, in the last resort, animistic thinking.

Reflecting on this tenacity of lingual haziness, we should first clear our minds from metaphoric, pseudo-concrete, magical and animistic thinking. No disaster "works" and an earthquake is nothing more than shock waves, never a fist that hits a city. The expression that "a disaster strikes" is as wrong as saying "the winds blow," because there is no separate process that swells the cheeks to blow. Wind is air in specific motion, not a separate being that makes the air move. In a conclusive way, it is the same with disasters: there is no distinction between a disaster and ("its") effects. Disasters do not cause effects. The effects are what we call a disaster.

For tracing the second line of argument, the findings of organizational sociology will contribute. To my knowledge, the most pointed results about this have been presented by Crozier and Friedberg. On a very abstract level they characterize an organization as a "coalition of human beings with the aim to solve their vital problems" (1979: 12). The capabilities of the organization for problem-solving, however, evolve step by step from those solutions that have been successful in the first place. Organizations, as Crozier and Friedberg put it, then tend to organize themselves around their success. In the long run, the successful solutions especially have to be defended against competitors and envy. Thus, the operation of organizations tends to turn into a hedgehog position. More and more, the operation of the organization is shifted into the center of the efforts of its members. At that point, an organization has turned to selfishness. Its only interests are self-preservation; the organization is administering itself, with the original reason for its establishment being turned into a subordinate, accessory matter.

Simultaneously, the perception of reality changes from a creative, problem-oriented awareness, toward a defensive, solution-oriented persistence. Instead of scanning for upcoming problems, the self-preserving organization defines reality within the framework of its available solutions. The interest is less in focusing on possible solutions for upcoming problems, but more on the applicability of the available solutions. This shift is important because it marks a difference in the ways of perceiving the world. The first way is analyzing the problem in favor of finding an adequate solution.

The second way is defining the problem according to the solution at hand. Thus, the latter is not focusing on reality, but cutting reality into the parts that fit into the organizational capabilities to handle them. Most definers of "disasters" act in the way that Prometheus used his bed. In the first place, their definitions of disaster do not focus on the vital problems of the victims, but on the solutions they have at hand or can provide. Reality then is exclusively seen from one approach; the solution defines the problem, and deductively, reality. "Disasters" are predominantly defined this way. The cases where warm clothing was sent to African famines, or thousands of tons of contraceptives or cough mixtures were sent to mass casualty situations are not only mistakes, but the logical outcome of the internal dynamics of self-preserving organizations.

Tracing the third line of argument, one may question whether the area named science underlies the same dynamics of perceiving the world and their problems and, if so, how disaster sociologists may conceptualize the problem named "disaster." Instead of criticizing the attempts of others, I will try to categorize some types of definitions favored in the field. The absolutely most frequent type is the *event* concept with the subcategories "time," "space," and "severity" (or a mixture of them). Next most frequent is the *stage or phase model concept*, which is often a variant of the event concept, but with an emphasis on a broader time scheme looking at the time and human activities before and after the event or the impact. A much simpler subvariant of the stage/phase type is the space-model that seems to be a derivation of the specific threats from bombing or explosions: zones of destruction, casualties, injuries, and rescue activities are drawn in concentric circles around the direct impact. The concept is typically used in disaster medicine and in emergency planning for nuclear accidents (see Suren 1982: 41; Notfallschutz 1986: 9). Also directly influenced by practice are those conceptualizations of disaster that are built along the typical planning and action schemes of emergency management. They use terms like "pre-emergency phase," "emergency-phase," "warning," "threat and evacuation stage," "dislocation stage," "relocation state," "post-emergency phase," and so on. I am sure that most readers will have identified Stoddard (1968: 11) behind the scheme just cited. He presented a wonderful table of the stage models used in the United States at the time he wrote his book.

A little different are those conceptualizations of disaster that use specific *ratios*: for example, the ratio between resources and demands. I will call these concepts the "lack-of-capacity" type. All of them define disaster as an agent much too fast, severe or overwhelming in relation to the capacities available. Thus, the disaster was too fast relative to the warning or too spacious relative to the rescue capacities, etc.

Another type of conceptualization is explicitly influenced by the natural sciences and technology. I will call these concepts the *systemic catalyst* type. Disasters are defined as the outcomes of misdirected energy or autodynamically colliding interactions between the components of complex systems. Turner (1978) and Perrow (1984), who can both be easily identified with these approaches, define disaster either as wrong amounts of energy at the wrong time and the wrong place, or as a self-induced resonance between technical subsystems leading to dangerous modulations and collisions. Both concepts have been transferred to organizational and human interaction to explain social failures as well.

Analyzing all these concepts, we have to ask about the type of definitions in use. At first glance, all examples cited seem to use real definitions, i.e. the definiendum is explicated with a specific definiens. For example, a disaster (*definiendum*) is an event (*genus proximum*) concentrated in time and space (*differentia specifica*). In terms of logic, the method is correct, although inappropriate in terms of being sociological. The question is how the *genus proximum* and the *differentia specifica* should look if it is a definite sociological *definiens*?

At this point, of course, no reformulation of sociology is intended. To avoid a misunderstanding, I am not talking about theoretical orientations, research designs or methodology. I am asking for the specifically social fabric of disaster. In other words: What is a disaster like if explicated in terms of human action? And if in trying to do so, will we find specific characteristics (*differentiae specificae*) which cannot be explicated in terms of human action?

MASKING REALITY WITH FALSE CAUSALITY

To myself, one of the most interesting results of analyzing the definitions categorized above is the implicit false causality. Looking at some examples again, the cases of false causality can be detected easily. In sociological terms, the "lack-of-capacity" type is the most revealing one. Most of those who have used this concept identify a specific shortage, but almost never describe it as a deficiency. Intentional or not, the shortage becomes masked by the turning of causality into false causality. The *event* was too sudden, instead of time was too short. Even if the lack of time is addressed, like saying that the warning period was too short, almost never will the warning system be systematically blamed as inadequate or insufficient. At most, the lack-of-capacity approaches suggest improvements in training, equipment, and resources. But in doing so, the risky criticism of this type of conceptualization comes to the fore. If this definition of disaster is radicalized, nothing else is a disaster but the lack of problem-solving capacities. Yet, that is exactly when we should become alert: Who is responsible for such a lacking? Instead of answering that question, it is much easier to turn causality toward the overwhelming forces coming from outside.

DISASTER AS COLLAPSE OF CULTURAL PROTECTIONS

At this stage of the argument, I will take time by the forelock to remind us of a disaster sociologist who never has found the appreciation he deserves. I am talking about Carr (1932), who was, as far as I know, the first in the field to try to understand disasters in terms of social action. I have certainly read his article on social change very selectively and I concede that my interpretation may miss the original intention.

Nevertheless, this idea was the nucleus around which this article was written. This effort, in fact, is dedicated to this disaster student. The idea that attracted my attention has not been developed systematically, yet it is evident in the following passage:

> Not every windstorm, earth-tremor, or rush of water is a catastrophe. A catastrophe is known by its works; that is, to say, by the occurrence of disaster. So long as the ship rides out the storm, so long as the city resists the earth-shocks, so long as the levees hold, there is no disaster. It is the collapse of the cultural protections that constitutes the disaster proper.
>
> (Carr 1932: 211)

Carr's conclusion signifies that disasters are the result of human activities, not of natural or supranatural forces. Disasters are simply the collapse of cultural protections; thus, they are principally man-made. Deductively, mankind is responsible for the consequences of his action as well as of his omissions.

As far as I can see, Carr interpreted the forces of nature as some sort of challenge (the "precipitating event"), that comes up during time (the "preliminary or prodromal period") (1932: 209). "In every disaster there is a preliminary period during which the forces which are to cause the ultimate collapse are getting under way" (ibid.: 211). If the cultural protections do not collapse under nature's attack, they have been proved to be functionally adequate, otherwise they are inadequate and collapse. To Carr, this collapse is the disaster proper, not the prior infight of nature and culture.

Albeit, there are some confusing inconsistencies in his argument. He often wrote that catastrophes "cause" disasters (1932: 210) or that a disaster "resulted from both earthquake and fire" (ibid.: 210), and he separates disasters and their consequences. Yet Carr has very clearly seen the problem of social causes and non-social causes that lead to disaster. His differentiation between disaster and catastrophe makes the point. Carr distinguished disasters not only on the basis of consequences, but also on the basis of (1) the character of the precipitating event, or catastrophe, and (2) the scope of the resulting cultural collapse (ibid.: 209). In his preceding argument, he explicated catastrophic change as: changes in the functional adequacy of cultural protections following catastrophes, i.e. the relatively sudden collapses of cultural protections resulting from catastrophes (ibid.: 207).

On the one hand, Carr realized the predominance of human action in the production of disasters. Thus, he knew that the collapse of cultural protections might be so rapid and radical that its functional adequacy may be totally questioned. Then, society in its entirety is endangered, not only its cultural protections (Clausen 1992). Therefore, both the extraordinary scope of cultural collapse and the extraordinary outburst of natural forces can lead to catastrophic change.

On the other hand, the helpfulness of Carr's differentiation must be doubted. In terms of logic, the introduction of extraordinary challenges does not alter the basic problem. If nature is too beastly or society too weak, and the loss of control too fast and complete, it is a human failure all the same. As a matter of fact, the breakdown of society or of some parts make a difference in terms of harm and damage, but not in logic. Both have collapsed because of their inadequacy. One inadequacy was a lack of foresight in seeing the challenges the protections will have to survive; another inadequacy was not taking additional precautions to allow keeping control even during very rapid and radical changes.

To my mind, Carr's attempt to describe disasters as social processes and as interrelated exchanges between natural and cultural forces was an important step in sociological disaster theory. Nevertheless, his concept of nature as a permanent source of trials, as a powerful adversary that continuously enforces acts of submission (1932: 209), reflects the contemporary conception of nature, instead of the inherent logic of his own approach. Thus, Carr turned causality the wrong way because of a wrong conception of nature. In his causal concatenation, a natural force would become a catastrophe the moment the cultural protections could not stand the challenge. Whether nature turns to catastrophe or not, can only be decided on the basis of cultural criteria and only the challenge of these criteria: Did the cultural protections collapse? If so, nature was catastrophic. Did they resist? Nature only tried to go berserk, yet conquered. If one remains consistent in terms of Carr's basic approach and its sociological and logical implications, a disaster is nothing else than the failure of protection measures, and that is the inadequacy of means (the cultural protections) in relation to given ends (to avoid their collapse).

FUTURE DISASTERS?

Seen that way, some major differentiations in the perception of "disaster" are necessary. The first problem is with the term "functional adequacy of cultural protection." If adequacy is decided after the extreme trial by natural or technological hazards, the question is how people will be able to create measurements in advance.

Logically, we know that knowledge in advance is impossible. Future modalities (possibilities) are only likely. Thus, the events that have not yet happened can only be anticipated in terms of probabilistic propositions, whereas for the events that have happened (the ex-post-facto-state) they are facts of the past. They can be documented definitely and completely (at least in principle). Consequently, two different conceptions of logic become necessary: one for the facts or things that have happened, and one for the events that are likely to happen in the future. Thus, the historiographic logic of facts has to be supplemented with a logic of probability.

Practically, humankind has to deal with future possibilities every day. Thus, the likelihood of the things to come has to be anticipated, which implies to think and to act under conditions of incomplete information and uncertainty. Generally, the mixture of the logic of facts and the logic of probability is not noticed. For most people, continuance is the reasonable stance and, during one's individual life cycle, only extremely few are forced to learn by personal experiences that the facts of today may be valid but useless to handle the tomorrow. Even in science the idea is predominant that future developments can be anticipated by extrapolating from the past and the present. In principle, this is the only way to deal with an unknown tomorrow (aside from more sophisticated mathematical and statistical methods). As a matter of fact, under conditions of definite and complete information, most developments will come true with a likelihood close to determination. On the other hand, man is not omniscient. In most cases, people decide under conditions of ambiguous and incomplete information and, knowingly or not, they bear the risk of failure. Altogether, very little is known about the totality of the world, the universe, the sum total of effects, and their interactions and the interferences they have already caused and will still cause.

Transferring these considerations to Carr's conception of disaster, "cultural protections" may be redefined as realizations of warnings, or, more precisely, of prognoses. The aim to take protective measurements necessarily presupposes the expectation of future failures. But what is the material these expectations are based on? On the one hand, they are based on the experiences which have been generated during the evolutionary process of trial and error (Murphy's law represents the highest generalization). On the other hand, they are based on substantiated imaginations, pictures of possibilities, visualizations, and visions. In most cases, it is a mixture of both, and in the modern secularized society it is called prognostication, forecast, or futurology. To tell the truth, nobody knows for sure how the latter differ from intuition, fortune-telling, soothsaying, prophecy, or revelation.

All this indicates the magic involved in anticipating the future. Yet, without extrasensory perception or supernatural guidance, the only way to look into the future is via probabilistic predictions. The difference from prophecies (such as those of Nostradamus) is very small and, correspondingly, the risk of falling for charlatanism is high. That is the reason why most people like predictions that are as close as possible to their known reality (or wishful thinking). Thus, every century has deciphered predictions and prophecies on the foil of common sense and empirical knowledge. An interpretation of the vision of doom by Nostradamus of the "holy fire that will fall from the sky" as nuclear explosion, would have been impossible before the Manhattan Project had been finished.

With all this in mind, I should like to turn back again to Carr's argument. Under certain circumstances, he had said, the forces of nature become a catastrophe. The test criterion for this change in state is the collapse of cultural protections. Consequently, nature becomes a catastrophe only when culture collapses. Thus, the prognosis of the forthcoming change of nature to catastrophe is verified after the collapse of cultural protections. In fact, a very strange syllogism! Nevertheless, for the present position, the individual errors in reasoning are not significant—except for one: What exactly might a meaningful criterion be like which marks the change in nature's state (from a "force" to a "catastrophe")? Keeping this clue in argument, I cannot find such a criterion in objective scales (such as the Richter scale for earthquakes) or in speed classifications for air currents (such as a scale from 1 to 12), because an earthquake in the desert or a heavy storm in the Arctic ice is meaningless (for our present state of mind at least). I do not exclude further wisdom and insight in the interrelation of human existence with the life of animals, plants, and matter. One day, perhaps, we will care about the whole planet; today, we are selfish to the point of annihilation. However, the argument I want to make is this: in contrast to the facts of history, future occurrences cannot be described in terms of a logic of acts and cannot be tested empirically. Thus, a decision between true or false is only deducible from an appropriate logic, the logic of probability.

Theoretically, an applicable test criterion for probabilistic propositions is available. To put it on a simplistic level, an applied test program for prognoses requires the complete and perfect duplication of our world on the basis of the algorithms that make it move and change. Similar to computer simulations, the selection of algorithms and the conceptualization of trial runs is a most delicate problem. As Perrow (1984) demonstrated, in most cases the failures that happened later in reality have not been anticipated in theory. The same problem is to be found in the field of technology assessment. What are the likely effects of a new technique (a chemical compound, a medicine, a product, etc.) and on which level of effects should the assessment process be stopped?

Of course, Turing (1936) gave a hint; the Turing generator makes it possible to transform every argument (variable) into algorithms. Nevertheless, the transformation requires not only a unique basis of calculation, but also a unique category of reference. Until now, a unique system of transferable references has not been developed. The attempts to do so (e.g. the transformation of everything into quantitative amounts of energy exchanges) are highly contested (see Rifkin 1980). Consequently, in the field of probabilistics no valid criterion is available that allows us to distinguish between true and false.

The lack of an empirical test criterion in the field of probabilistics (mis)leads for the reuse of even those arguments which are refutable or which have been refuted with the help of empirical facts. In the end, the transfer of arguments out of the system of the logic of facts into the system of the logic of probability irrationalizes every argument. Without a test criterion, every argument will become equivalent in the sense of indifference, because true and false become undecidable.

In the sphere of social interaction, indifferent and undecidable conditions are hardly bearable. Thus, many sociologists suppose that the most important objective of social action is to get control over the conduct of others and of nature (see Burns 1958; Elias 1983). To avoid dangerous surprises and uncertainties, social action is preferably transformed into reliable repetition and certitude. Consequently, perpetual action is often firmly established by rituals, customs, norms, institutions, or organizations, which react upon human action like a silent but unchangeable force of circumstances. Accordingly, human action appears in process and in manifestation; both forms will influence interaction as counterparts.

The most efficient way to get control over complex systems and interaction is to manipulate their lodestar, or, in other words, the algorithm with the highest complexity. In religious systems, it might be the deity's will revealed by the priesthood. In feudal systems, it might be the fief, distributed by the nobility, and, in modern times, it might be the status based on the money that people have available. The advantage of controls by the top algorithm is the extreme efficiency: the alteration of one factor alters the whole system.

On the other hand, there is the risk that the subordinated algorithms will win an extensive autonomy. Similar to the rise of the city during feudalism, this autonomy may steadily undermine the efficacy of the top algorithm. On the surface, the whole system seems to be under the lodestar control, but underneath contraproductive effects thwart the original intentions. In most cases, the collisions of different intentions are not realized as the outcome of social action or social change, because people mostly do not know the true intention of others. Thus, the collision of different intentions and planning is more often interpreted as a disaster than as loss of control.

Speaking more generally, the explication of disaster as an unplanned and unintended result of human activities, which is a counterstroke to the planned and intended effects of action above a certain level of tolerable disturbance, can lead to a misleading sociologism. It is not only human interaction itself, or interaction with material culture and its autodynamics, that may generate failures, but also the interaction with nature and its own autodynamic and self-organizing processes. Most authors in the field (including Carr) have reflected on nature's influence on human interactions, but only a few have analyzed the significance of nature's autodynamics without treating nature animalistically or in categories of an acting subject. Nevertheless, as a living system, nature is interacting with the effects of human metabolism. Thus, the human effects of first (planned and intended effects) and second order (unplanned and unintended effects) may also collide with the unforeseen response of the autodynamics of nature. Altogether, the chance to fail is increasing exponentially, because most effects of human interferences in natural and cultural processes are unknown.

From a statistical perspective, it must be foolishness or madness to intervene in systems that people depend on without knowing how the systems work and how they will react. Above a certain quality of intervention, the chance to destroy the basis of living becomes possible (and, in this respect, Chernobyl is a "good" proof). The dilemma we have to deal with is this: without the knowledge of the effects of our action, and without the knowledge of the functioning of all the systems that are interfered with, the resulting risk of failures becomes very high. But without interferences, experience and knowledge is impossible. Consequently, the most important parts of our journey of discovery are the failures, because only failures will unlock the secrets of the universe of the unknown. As in the philosophy of science, which does not accept verification as final proof, human praxis should not accept success as final proof, because one never knows whether it was a lucky chance or proper application of applied knowledge.

Therefore, from my perspective, correct praxis is the key word in human action, but this cannot be completely defined in terms of technological success or of the correctness of the planned and intended action. As long as the unplanned and unintended effects of human action and of the autodynamics of nature are not added to our concept of reality, we only believe in metaphysics, but not in rational knowledge. Separated from a definite theory of the entire interplay of effects of all kinds, every epistemology remains pseudo-concrete.

From a very abstract level, regarding the investigation of disasters (failures) starting after their occurrence, resembles the inductive method. From a unique and single "event" a universe of possible causes has to be concluded. Yet, without the imagination of this universe, of the totality of effects, the range and scope of possible causes cannot be anticipated. Mere description or vertigo in the circle of hermeneutics will be the alternative. Nevertheless, inductive disaster research is the step-by-step method to explore the totality of effects. Comparing the intended with the unintended, we will be able to detect the algorithms that lead to disaster at the same time. Vice versa, we will find out how we have to use the algorithms to avoid disasters. Seen that way, disasters are the only falsifications we have to prove the truth, i.e. the empirical correctness of our theories. Moreover, disasters are the proof of the knowledge of our knowledge. They are the key algorithm of our epistemologies. In this sense, disasters are fairly well sociological.

4

DISASTER AS SYSTEMIC EVENT AND SOCIAL CATALYST

Gary A. Kreps

INTRODUCTION

I see more agreement than disagreement among social scientists about what disasters are and how they can be distinguished from other types of societal phenomena (cf. Hewitt 1983b). Thus instead of criticizing the views of others, I will characterize as pointedly as possible the consensus that I believe exists. Sometimes explicitly and sometimes not, disasters have been interpreted for over three decades as *systemic events* and *social catalysts*. This means that certain kinds of historical circumstances fit neatly within the boundaries of disaster research, while others do not. Such exclusiveness is a good thing because the boundaries of disaster research remain appropriately broad. They include a wide range of environmental, technical, and sociopolitical events which should be compared more systematically than has been the case in the past.

What follows amounts to the continuing clarification of disasters as *social constructions*. That clarification sensitizes disaster researchers to life history perspectives on actual or potential events. The events themselves involve social definitions of physical harm and disruption of routine activities in societies or their larger subsystems. The first section of the paper provides a formal conception of disasters that builds on ideas expressed by Fritz (1961), Dubin (1978), and Barton (1989). The second section of the paper illustrates this conception with life history studies of the Mexico City earthquake (1985) and Chernobyl nuclear power plant accident (1986). The third section shows how these two events can readily be compared using the conceptual tools provided earlier. The paper closes with a brief on how sociological knowledge should advance within the exclusive but broad boundaries of disaster research.

GARY A. KREPS

DISASTER AS SYSTEMIC EVENT AND SOCIAL CATALYST

> A disaster is an *event* concentrated in time and space, in which a society or one of its subdivisions undergoes physical harm and social disruption, such that all or some essential functions of the society or subdivision are impaired.
>
> (Paraphrased from Fritz 1961: 655)

> There does not seem to be any compelling reason why we may not think of *social catalysts* and use them in our theories. For example, in the study of social behavior under conditions of disaster, the event serves as a catalyst whose presence is necessary for examining the social and psychological concepts of interest to disaster specialists. It makes no difference whether the event studied is a flood, an earthquake, an explosion, or whatever.
>
> (Paraphrased from Dubin 1978: 115–116)

> If one starts combining physical-temporal characteristics and social definitions, somewhere in a list of about fifteen permutations and combinations of these characteristics, one finds community level physical disasters. That is the location of most disaster research. There are also regional physical disasters, which tend not to have been studied because they have been very rare in the United States since funding for disaster research came through.
>
> (Paraphrased from Barton 1989: 348).

I have paraphrased the above ideas from Fritz, Dubin, and Barton to illustrate the historical point that sociology has played a major role in clarifying the meaning of disaster. For some time now, sociologists have interpreted disasters as special types of societal phenomena, in part because they are dramatic historical happenings (events), and also because they compel collective reactions (social catalysts). By recognizing these key conceptual features of disasters, we are able to see how and why they can be distinguished from environmental hazards of various types, and an even broader range of societal concerns.

Serious attempts to define disasters are usually traced to the above definition by Fritz (1961).[1] That definition was grounded primarily in the strategic bombing studies of World War II, as well as numerous peacetime disaster studies undertaken by the National Opinion Research Center (NORC) and later the National Academy of Sciences–National Research Council (NAS–NRC). Cited repeatedly over the past three decades, Fritz's definition has stood the test of time very well. It balances attention to physical harm *and* social disruption, and it implies that many different kinds of events involve social definitions of these unfavorable conditions (Barton 1969, 1989).

Having reached at least implicit agreement over the years about what a disaster is sociologically, many researchers have been preoccupied with identifying and illustrating its key dimensions (Dynes 1974; Mileti, Drabek, and Haas 1975; Quarantelli and Dynes 1977; Burton, Kates, and White 1978; Perry 1982; Kreps 1984, 1989b; Drabek 1986, 1989a; Britton 1987; Kroll-Smith and Couch 1991). Some researchers have examined how social systems react to physical harm and social disruption *after* an event has occurred. Others have examined what social systems do to increase or mitigate the risk of physical harm and social disruption *before* an event has occurred. But in either case, disasters are thought to be implicit or explicit catalysts for collective action.

While precise thresholds of when historical happenings are socially defined as disasters have never been determined (see Wright and Rossi 1981), no one denies that such thresholds exist. There is no argument, for example, about whether the Bangladesh cyclone (1970) which killed 200,000 people, the Managua earthquake (1972) which killed 25,000 people, or the Chinese earthquake (1976) which killed 250,000 people qualify as disasters. And when technological accidents such as the release of a toxic chemical (methyl isocyanate) at Bhopal (1985) or the breech of containment at Chernobyl (1986) occur, no one questions their designation as disasters either. And at least some disaster researchers include political strife such as that which has occurred in Nigeria (Biafra), Pakistan (Bangladesh), Ethiopia, Lebanon, Somalia, and Yugoslavia as disasters. The most obvious characteristics of these and other events are as follows: *they involve considerable harm to the physical and social environment; they happen suddenly or are socially defined as having reached one or more acute stages; and something can be done to mitigate their effects before or after they happen* (Erikson 1976).

Assessing the severity of physical harm and social disruption of events such as the above must always be based on what is identified as the impacted social unit. Thus natural disasters are common in most societies, and their effects are absorbed readily in large and technologically advanced ones. But not all societies are large and technologically advanced; and even if they are, disasters become less regular and their impact ratios become much higher as one moves from the societal to the regional, community, and household levels. Recovery obviously takes on different meaning depending on which societal level is being considered (Bates and Peacock 1987). And while disaster research historically has confined itself to the societal level or below, societies themselves can be seen as parts of a more inclusive world system that is equally disaster prone. Thus wars are common occurrences at the global level, the planet has survived, but there are major uncertainties about a massive nuclear war.

Potential disasters (or hazards if you prefer) are just as important (practically as well as theoretically) as those which actually occur (Britton 1987). Imagine, for example, a major earthquake in the heart of San Francisco during working hours, as opposed to one outside the central city during the evening (such as the Loma Prieta earthquake which occurred during the 1989 World Series). Now place yourself in the role of either a public official or resident of an earthquake-prone area. With respect to disaster prevention and mitigation, which of these two scenarios should be considered? The answer is that both of them are circumstances that might have to be confronted. The simple point is that any conception of disaster implies worst case scenarios that are sociologically credible.

A life history perspective is essential for studying disasters because they are social constructions (Hewitt 1983b; Quarantelli 1987b, 1989a). This means that any social system vulnerable to disasters should be examined before as well as after an event occurs (Shrivastava 1987; Drabek 1989a; Perry 1989b). With that in mind, I offer only a slight modification of Fritz's definition of disasters. Disasters are:

> nonroutine events in societies or their larger subsystems (e.g. regions, communities) that involve social disruption **and** physical harm. Among, the key defining properties of such events are (1) length of forewarning, (2) magnitude of impact, (3) scope of impact, and (4) duration of impact.

Thus, disasters have life histories which can be designated in time and space. The phrase *nonroutine events* distinguishes disasters as unusual and dramatic happenings from the reservoir of everyday problems and concerns which confront humankind. The conjunction **and** in bold print distinguishes disasters from emergencies. The designation *societies or their larger subsystems* means that social and physical effects, however they may be measured, must be socially defined as such at the community level or above (Bates and Peacock 1987). Disaster therefore serves as a sensitizing concept (Blumer 1969; Dubin 1978). It calls attention to a certain class of historical circumstances and highlights several underlying dimensions.

There are four underlying dimensions that sociologists have given much attention to in thinking about disasters as events. *Length of forewarning* is the amount of time between the identification of hazardous conditions and the actual onset of effects on particular locations. The warning time can be very short to quite long. *Magnitude of impact* refers to the severity of social disruption and physical harm. Here lower to higher severity is socially-defined disruption of normal routines, damages to natural or built environments, and direct effects on human beings (deaths, injuries, and illnesses). *Scope of impact* refers to the social and geographic boundaries of social disruption and physical harm. These boundaries can be highly localized to quite diffuse. Finally, *duration of impact* is the time lag between the onset of social disruption and physical harm and when the disaster is no longer defined as producing these effects. That time frame can be short to quite open-ended.

The above conception implies that event characteristics are independent variables while individual and social responses, however defined and measured, are dependent variables. Certainly this causal direction has been emphasized within the disaster research legacy. But one can just as easily reverse the causal argument, as would be the case in studies of disaster vulnerability, mitigation, or preparedness. In either case, the historical characterization of disasters by sociologists as nonroutine, dramatic, and system-threatening events sets important boundaries on the kinds of phenomena that are being considered (Bates and Peacock 1987, 1993).

The fact that thousands of people die on American highways in any given year does not mean that a disaster has been socially determined. The crashing of a fully-loaded commercial airliner in the heart of a central business district is certainly more likely to be defined as a disaster. The nuclear power plant accident at Three Mile Island was defined as an emergency, and was certainly a potential disaster. But potentiality became social reality at Chernobyl. Poverty, hunger, and social unrest are chronic societal concerns. Economic depressions, famines, and wars are interpreted as disasters. Global warming and ozone depletion are environmental hazards. The possible disastrous consequences of these planetary threats are matters of considerable scientific and public debate. A nuclear attack is a reality for which there is a historical precedent. A nuclear war is a possibility that can only be postulated. But postulated or not, disasters are conjunctions of historical happenings and social definitions of physical harm and social disruption.

Studying disasters also calls for a comparative perspective. The range of events typically studied by sociologists include most natural hazards (e.g. tornadoes, floods, hurricanes, earthquakes, volcanic eruptions) as well as accidents (e.g. major air disasters, explosions, dam breaks, large-scale fires) that involve comparable physical destruction and social disruption. Far less frequently studied events include famines, epidemics, economic depressions, political revolutions, and wars. Interestingly enough, wartime and peacetime nuclear events have always held a special place in the disaster research legacy. Prince's (1920) research on the Halifax shipwreck and resulting massive explosion of trinitrotoluene (TNT) is generally considered to be the first social science study of disaster. However, contemporary disaster research is better dated to the strategic bombing surveys of World War II, which included studies of Hiroshima and Nagasaki. Many subsequent peacetime disaster studies in the 1950s and 60s were done under civil defense auspices. The funding rationale was as follows: peacetime disasters include many physical impacts that are comparable to nuclear weapons effects. These events can therefore serve as natural laboratories for observing how individuals and communities cope with conditions of acute stress. That rationale has been debated continuously over the years.

Does the preceding discussion mean that disasters are more important than other social phenomena? Of course not. Are they different sociologically?

Most certainly. And that is the key to understanding disasters on their own terms, and then comparing them with other societal concerns (Hewitt 1983b). As identifiable events, disasters include actual or potential impacts on societies or their larger subdivisions. The list of disasters may be quite long; its length depends on how one defines the boundaries of the field. I would keep these boundaries broad to include environmental, technological, and sociopolitical events. Before and/or after events occur, social systems take actions that are related in one way or another to them.

Two case studies of disasters as systemic events and social catalysts

The first event, an earthquake, typifies what Barton (1989: 346–351) refers to as *community physical disasters*. The effects of such disasters are fairly concentrated, the events themselves are defined by the masses and elites alike as unusual happenings, and while physical destruction and social disruption are severe, the ratios of damage to remaining resources are not overwhelming. The second event, a power plant accident, typifies what Barton refers to as *regional physical disasters* because their effects are much more widespread. In this summarized case, entire nations that played no decision-making role in the design or operation of the plant were impacted by the accident. The most obvious difference between the power plant accident and the earthquake is that its regional physical impacts included exposure to radioactive contamination. In reviewing these cases, notice the importance of a life history perspective for a sociological interpretation of what happened.

Mexico City earthquake, 19 September 1985[2]

Mexico City was (and remains) a community of enormous size and importance. With a population at the time of the earthquake estimated at between 18 and 20 million residents, this sprawling urbanized area was becoming the largest metropolis on the planet. Over 20 percent of the Mexican people lived in a city that produced almost half of the Mexican gross national product. It absorbed about a quarter of the federal budget and about a third of all government investment. It was both the political capital of the nation and its historical, economic, social, and cultural center. Having become a world city in the latter half of the twentieth century, Mexico City evidenced all of the vitality and major problems conventionally associated with rapid urban growth.

The Mexico City area experiences numerous seismic shakings every year, on average about 90 per year that register 4 or more on the Richter scale. In fact, in the first 45 days following the earthquake reported on here there were at least 150 tremblers that ranged from 3.5 to 5 on the scale. The first

Mexican earthquake recorded in official chronicles occurred in 1637, although there are reports of earthquakes in that country as far back as 1460. Major earthquakes have occurred six times in the last 150 years (1845, 1859, 1911, 1932, 1957, and the one in 1985). The most recent one in July 1957 had its epicenter about 170 miles away from Mexico City (on the Pacific coast). It killed about 160 people, damaged several thousand buildings, and resulted in property losses of about 25 million dollars (mostly in the downtown area of the city).

The metropolitan area is especially vulnerable to earthquakes for two very important reasons. First, Mexico City was a town built during the Aztec era above Lake Texcoco, which was a body of water existing until the beginning of the twentieth century. As a result, many parts of the metropolitan area, and especially the old city in the central zone, are resting on very insecure ground. Seismic movement in such ground registers five to six times greater than that occurring on rocky ground on the periphery of the urbanized area. Second, the size of Mexico City's population has grown dramatically in the last twenty years, close to tripling its seven million resident base of 1970. A much larger number of people are therefore at risk, a substantial proportion of whom are poor.

While seriously and repetitively threatened by earthquakes, the level of disaster planning at federal and metropolitan levels was quite limited prior to September 1985. The federal government did have a plan for disasters generally, one which assigned responsibility for coordinating emergency response to the Mexican military. Clearly different from the United States, upon declaration by the President the army was to assume the lead role for all response actions. But, within the federal district of Mexico City itself, there had been virtually no formal planning for disasters of any kind, and certainly not for an event of the severity encountered in September 1985. Certain federal departments and units, such as those within public works and the subway system, had standard operating procedures and some plans for handling disruptions of their own operations. But there was no overall, system-wide planning at the federal district level. At the metropolitan level, a civil protection unit had been created in 1984 under the mayor's office. But the unit was very small, with few resources and little or no credibility.

An earthquake occurred without warning on 19 September 1985 and a major aftershock followed about 36 hours later. The earthquake measured 8.1 on the Richter scale, making it one of the worst in the nation's history. The aftershock measured 7.5 on the scale, thus aggravating the effects of the initial event. While the impacts of the earthquake and aftershock were regional in scope, the most severe damage occurred in Mexico City and several neighboring communities. Although unofficial estimates were much higher, final official figures indicated that 4,287 people were killed, 5,748 suffered serious injuries, and 10,188 suffered minor injuries. A total of 4,096 persons were listed as rescued alive from the ruins of buildings.

Official figures also indicate that 5,728 buildings were damaged or lost in the metropolitan area. Of this figure, about 15 percent totally or partially collapsed at the time of the earthquake, 38 percent suffered structural damage, and 47 percent suffered nonstructural damage. A large number of these buildings provided multi-family housing. Although difficult to determine precisely, the number of people suffering damage to their dwellings and possessions was certainly as high as several hundred thousand, and perhaps as high as one million.

Virtually all sectors of Mexico City's economy were affected, including health, education, telecommunications, manufacturing, public services, tourism, and housing. Effects on the housing sector were perhaps the most obvious. Close to 70 percent of the collapsed or damaged buildings were in this category, and the above estimates of damage do not include private buildings that had been damaged and repaired by owners. The most dramatic losses in the housing area were those involving large high rise developments. Loss of life was heavy here. Considerable numbers of low-rise units in the historic center of the city were also lost, but with far fewer deaths. Much of the destroyed or damaged housing was old, overcrowded, and in bad condition prior to the disaster. As is typically the case in most disasters, survey data suggest that the people of Mexico City most seriously affected by the earthquake were those with low to moderate incomes.

While housing was seriously affected, there was proportionally greater damage to health services and education. A total of 4,260 hospital beds, or 25 percent of existing capacity, were listed as lost; and a total of 761 public school buildings, or 30 percent of existing stock, were listed as damaged to varying degrees. Of these school buildings, 15 collapsed or had to be demolished, 310 required major structural repairs, and 436 required minor repairs. Of 1,044 registered private schools, nine had to be demolished, and 904 had minor to more serious damages. The loss of government offices was also considerable. The heavy damages to hospitals, schools, and government offices was in no small way the result of their heavy concentration in the central areas of the city.

Public utilities such as water, electricity, telephone, and transportation suffered varying degrees of damage as well. Water service to over 400,000 residents of the city was disrupted. As a result of damages to eight power generating substations and about 1,280,000 electrical installations, electricity services had to be suspended in many areas to allow for repairs and to lessen the possibility of fires and explosions. The partial collapse of the central telephone building was a major problem. Apart from disrupting local telephone communication, almost all national and international long-distance, telex, and telegram lines were rendered inoperable for several days. Fortunately, there was no significant damage to the city's highways, railways, ports, or airports. Bus services were disrupted in about 40 percent of all routes due to debris-clogged streets, but generally full service was restored fairly quickly. The subway system suffered very little damage and was operating close to full capacity within six hours of the quake.

Given this description of deaths, injuries, and property damage, it is hard to argue with the conventional characterization of the earthquake as an important sociological event. The largest urban complex in the world had been harmed in many tangible and perhaps intangible ways. But although the physical impact and social disruption of the earthquake were severe (and socially defined as such), and despite the fact that there had been very little formal disaster planning, the capability of the community to respond remained largely intact. This is because the proportionality of physical harm relative to remaining human and material resources was low.

How then did Mexico City react to the disaster? The community's emergency response mirrored its pre-disaster social structure. That structure was immensely large and organizationally complex, ostensibly controlled by the federal government, but relatively autonomous in providing basic public services. Thus, it is not surprising that the federal emergency plan was activated within hours after the quake, with the military being given the lead role in responding to the disaster. But the federal plan was implemented only in localities outside the capital. Inside the capital, the role of the military was restricted largely to providing security and crowd control at search and rescue sites.

The organizational response of public and private agencies in Mexico City was massive, decentralized, and highly complex. Within the public sector, agencies and departments from the national, federal district, and 16 subdivision levels quickly became involved in the restoration of public services. In addition, a wide variety of private agencies, businesses, voluntary organizations, and ad hoc citizen groups launched major activities. Concurrently, millions of individual and group volunteers mobilized to meet immediate demands for search and rescue, care of casualties, and aid to those whose housing was damaged or destroyed. Eventual responsibility for overall coordination of this massive emergency response was placed in the mayor's office. But, in the absence of prior planning, that coordination did not develop for several days, and it was necessarily improvised as conditions dictated during the remainder of the emergency period.

In effect, coordination developed gradually during the emergency period from the sheer momentum of highly diffuse human activities. While considerable confusion and some conflicts about responsibility and authority occurred, key emergency needs were met in a rapid manner. As might be predicted from the pre-disaster social structure, the 16 major subdivisions of Mexico City operated more or less independently, with those more severely impacted receiving help from those that were less so. The major problems facing responding groups and organizations during the first several days included: formal search and rescue (the vast majority of those extracted from the rubble were rescued informally by those immediately on the scene);

delivery of emergency medical care; provision of food, water, and emergency shelter; the carrying out of more systematic damage assessment; maintenance of security and traffic control; and handling of the dead. With the damage spread across a wide area of the city and hundreds of sites requiring attention, much of the organized response was tied to specific areas and not centrally controlled.

The restoration of public services illustrates nicely the organizational character of the emergency period. As noted above, damages to power generation, communications, and water systems were severe, but there was no major impact on the transportation system. The activities of lifeline organizations during the emergency period had a number of common characteristics. First, the response was massive and multi-jurisdictional. While some coordination of activities occurred among lifeline agencies that interacted routinely prior to the disaster, during the first several days many agencies acted autonomously. Pockets of coordination then emerged among units engaged in common activities. Second, the initial response of all organizations focused on damage assessment, information gathering, and rapid restoration of services. Third, those lifeline organizations that were less seriously affected extended emergency activities into geographic and functional areas that were not part of their traditional domains. This extension of activities was particularly pronounced for public works and transportation agencies. With vast amounts of human and material resources available, then, essential public services were restored quite rapidly, albeit not always efficiently, within the first several days following the quake.

Similarly illustrative of the community's emergency response were the activities of PEMEX, the national petroleum company of Mexico. Employing thousands of people, the company suffered no major damage from the earthquake to its operations or equipment. However, PEMEX assumed a major role during the immediate emergency period. It was able to mobilize 5,000 workers and engage in a variety of emergency tasks, particularly those related to search and rescue, debris clearance, and sheltering. With respect to the latter, and working with the Union of Petroleum Workers, PEMEX established and managed two large emergency shelters (a college building and a refinery facility). In undertaking these activities, the company maintained great autonomy and limited its contacts with other organizations to its own union and the military.

At first company officials felt that they were not authorized to handle such unconventional domains as search and rescue and sheltering. But within two to three hours of impact the decision was made to mobilize PEMEX resources. An emergency operations center was established at PEMEX headquarters and emergency communications equipment was installed. A map of damaged areas was constructed, the city was divided into four sectors, and company officials decided to focus its rescue activities on 24 major sites. The sites included such important buildings as two heavily damaged hospitals. Within each sector, one PEMEX supervisor was placed in charge of field operations. Each sector also had one coordinator in charge of the number of sites at which PEMEX worked in search and rescue and debris clearance. Operating independently, PEMEX extracted over 1,000 bodies from the rubble and reportedly rescued over 400 living persons.

The massive mobilization of local resources, combined with a major infusion of foreign assistance, resulted in an effective, if not always efficient, emergency response. Evidence of that effectiveness was the largely favorable evaluation of all aspects of the emergency response by two random samples of the adult population of Mexico City. With respect to long-term recovery, the major problem that had to be addressed was restoration of residential housing. Survey data suggests that perhaps as many as 250,000 residential units were damaged to varying degrees. But within two years almost 100,000 victim families had their housing restored or replaced through a variety of governmental and nongovernmental housing programs.

A key organizational mechanism for the restoration of housing was the creation of a federal agency to manage what was termed the Housing Reconstruction Program (HRP). More than 48,000 dwelling units were built, rebuilt, or repaired by this agency alone in eighteen months. The project was financed through equal participation of the World Bank and the Mexican government at a cost of over 373 million dollars. The Mexican Red Cross, in collaboration with the American and several other national Red Cross societies, was a major actor in housing rehabilitation as well. Beneficiaries of the housing program had to repay only the direct costs of reconstruction, with monthly payments based on a percentage of minimum wage. More than 1,200 small and medium-sized private companies worked in the program. About 175,000 jobs were created by the program, including 1,200 related to the HRP agency itself. Interestingly enough, the success of the program generated increased public demand for low cost and demonstrably higher quality housing.

Finally, there has been no systematic data on economic and social indicators reported which would suggest that the earthquake had severe long-term consequences for Mexico City as a whole. While perhaps up to 10 percent of the population was harmed directly in some fashion, and most other residents faced temporary economic and social disruption, a compelling argument can be made for the adaptiveness and resiliency of this world city. Mexico City's economic and social structure remained largely consistent with historical trends and, if anything, its political structure became more, rather than less, legitimated as a result of actions taken in response to the earthquake.

Chernobyl nuclear power plant accident, 26 April 1986[3]

At the time of the accident, Chernobyl was a small town of 12,500 people in the Ukrainian republic of the former Soviet Union, located about 100 kilometers from Kiev, a city with over two million inhabitants, which was the capital and urban core of the region. Containing mostly marshlands, the Chernobyl area was not a major agricultural region. The agriculture that did exist was primarily dairy farming. However, the rivers and lakes of the Chernobyl area made it both attractive for summer vacations and an important source of water for Kiev and the entire Ukraine. Located 15 kilometers to the northwest of Chernobyl was a large nuclear power station with the same name. But in the years preceding the April 1986 accident, the people of Chernobyl had little knowledge about or direct contact with this facility.

The Chernobyl power station had four RBMK Soviet nuclear reactors in full operation at the time of the accident, with two more under construction. The initials RBMK are a Russian acronym meaning "great power channel type" reactors with nominal power exceeding 1,000 megawatts. Such reactors are cooled by water and moderated by graphite. The construction of the Chernobyl power station began in 1971, and by 1983 its four operating reactors had an installed power capacity of 4 million kilowatts. Three kilometers away from the power station was the city of Prypyat, with a population of nearly 50,000 people. This young and rapidly growing town was the prime residence for Chernobyl power plant construction workers and operators.

The Soviet Union had developed a strong and growing civilian nuclear power program by April 1986. While the country depended on nuclear power for only about 11 percent of its electricity production (as compared to 70 percent for France, 60 percent for Belgium, 43 percent for Sweden, 40 percent for Switzerland, and 31 percent for West Germany), in so doing it was generating about 10 percent of the world's nuclear power from 43 operating reactors. The electricity capacity of these 43 reactors was 27 billion watts. Under construction were another 34 reactors, or potentially 37 billion additional watts, and under planning were 34 more reactors which could produce about 36 billion watts of electricity. Soviet plans indicated a major shift in future plants away from RBMK-type reactors and toward what some nuclear experts consider to be more modern and economical pressure-vessel type reactors.

The basic fuel of nuclear reactors is uranium. The technology for producing electricity calls for small particles called neurons to be used for splitting up uranium atoms, and this produces heat and more neurons, which keeps a chain reaction going. The speed of the neurons, which must be moderated for the chain reaction to work, are slowed down by graphite in the RBMK-type reactors. The power of a nuclear reactor is controlled by changing the number of neurons: the power is increased if more neurons are produced than used up; it remains steady if the number produced matches the number used up; and the reactor eventually shuts down if more neurons are used than produced. The heat produced by the chain reaction turns water into steam, and the energy of the hot steam spins a turbine, which in turn spins an electric generator.

The hot steam must be kept in a strong container because it is maintained at high pressure. In the Chernobyl RBMK-type reactors the strong container consists of about 1,600 small pipes, called pressure tubes, which hold the uranium and hot water. An important aspect of the four Chernobyl RBMK-type reactors is that they have only partial containment of the whole reactor system. At the time of the accident, containment was generally confined to the presence of strong leak-tight buildings surrounding the pipes which hold the hot water and uranium fuel. The steam pipes below the reactor cores of Chernobyl reactors were inside leak-tight boxes that were connected to a huge pool of water under the entire building. If one or more of the pipes in the boxes broke, radioactive particles would be channeled to and trapped in the water, and the containment boxes would hold. However, the steam pipes above the reactor cores were inside ordinary industrial buildings. Thus, if any of the pipes broke above a reactor core, a release of radioactive steam would occur. Any subsequent release of radioactive contamination would then depend on how effective other emergency systems (i.e. shutdown of the reactor and emergency core cooling) were in preventing damage to the uranium fuel.

The accident began as an efficiency test being carried out on a turbogenerator at the time of a regularly scheduled shutdown (for routine maintenance) of the Unit 4 Chernobyl reactor. In order to understand the test, suppose that a nuclear reactor is shutdown immediately following an accident. This would mean that the reactor could not generate its own electric power directly. Suppose also a simultaneous loss of external electricity to run essential reactor safety systems in the event of an accident. Under this rare, but certainly possible, circumstance a diesel generator is supposed to kick in. However, it takes approximately 20–50 seconds for that to happen. The purpose of the test was to determine the capacity of the Unit 4 turbogenerator to supply uninterrupted electricity (from residual mechanical energy) during these 20–50 seconds. A successful test during a regular shutdown would increase management confidence that, should a simultaneous shutdown and electrical failure occur, the emergency core cooling and other safety systems would remain operational until standby diesel generators could supply emergency power.

A combination of inadequate standard documentation (i.e. operator manuals and safety standards), human error during the efficiency test, and the basic design of the RBMK-type reactor led to the worst nuclear accident in world history on 26 April 1986 at about 1.30 a.m. The immediate cause of the accident was an unintended power surge which led to a sudden heating of the uranium fuel. The heated fuel broke up into many small pieces. The heat from these pieces caused a rapid boiling of the cooling water in the pressure tubes and, as a result, a number of these tubes burst. Radioactive steam then escaped from the pressure tubes, burst the metal housing around the graphite, and lifted the concrete shield on top of the reactor. All the remaining pressure tubes then burst open. The power surge not only destroyed the top half of the reactor core, but, in the absence of complete containment (as noted above), it also destroyed the building immediately above the reactor and some of the walls on either side of the building. Burning fragments of radioactive fuel and graphite were thrown out as well, causing about thirty fires, primarily in Unit 4, an adjacent turbine building, and the roof of Unit 3. Major damage was confined to the Unit 4 reactor. Units 1–3 kept generating electricity for 4–22 hours, by which time they had all been shut down because of the increasing radioactive contamination of the entire plant complex.

Putting out the fires was the initial concern because for a time the Unit 3 reactor was threatened as well. Fire units from Prypyat and Chernobyl responded immediately, and the fires were extinguished by about 5.00 a.m. or about three-and-a-half hours after they started. However, the major problem was that the destruction of the Unit 4 reactor and containment had led to a substantial release of radioactive contamination for about nine days. That contamination had to be monitored on the ground and in the air with radiation detection equipment, and it had to be controlled with any and every means available.

The release of fission products did not occur as a single explosion at the plant. Initially, there was an intense release associated with the burning of graphite from the Unit 4 reactor. Large amounts (about 5,000 metric tons) of boron, dolomite, sand, clay, and lead were dropped on the reactor to suppress this release of fission materials. Earth moving equipment and helicopters were used extensively to complete this task. Release rates fell over the next three to five days as a result of this suppression activity. The rates then began to increase for the next three to four days as the materials that had been dropped insulated the reactor core, which then heated up again. But by the ninth day the heat loss was exceeding the heat-up rate of the smoldering graphite. Release rates declined more or less continuously from that day on.

Initial attempts to monitor and control radioactive contamination were largely improvised, and grave personal danger faced anyone working in the power plant complex. Simply put, being on site or otherwise engaged in initial damage control activities was likely to result in acute radiation sickness. It is not clear how many power plant operators, construction workers, fire fighters, and military personnel were involved initially in controlling fires and suppressing fission products. Nor is it clear how many of these individuals were aware of what radioactive contamination could do to them. But it is very clear that the medical effects for those exposed to radioactive contamination during the first 24–48 hours were serious to lethal. Thus, the widespread characterization of these early responders as heroic seems very appropriate.

It is highly likely that at least several hundred people were working on site during the first hours after the accident, and within a few days many thousands of miners, transportation workers, firemen, and military personnel (from Kiev, Moscow, and Donbass) were mobilized. At minimum, 300–500 people were admitted to Soviet hospitals during the first several days with an initial diagnosis of acute radiation sickness. This diagnosis was reportedly confirmed after clinical, instrumental, and laboratory tests for 237 individuals. The mildest cases of acute radiation sickness received up to 200 rems of irradiation, a very harmful level. The most severe cases received from 600 to 1,600 rems, which generally results in death. The number of people officially reported as killed from, or immediately after, the accident was 25 power plant personnel and six firefighters who were unprotected against lethal dozes of irradiation. Those suffering from acute radiation sickness suffered damage to their bone marrow, blood, skin, connective tissue, stomach, and intestines. Their treatment required the services of various Soviet specialists, and marrow transplant specialists assisted Soviet doctors in thirty marrow transplant operations.

While acute radiation sickness was an immediate threat to people on site, the longer-term medical effects of radioactive contamination was of critical concern as well. This is because radioactive particles can travel great distances from their source, the health impacts of which are based on cumulative exposure. About 50 percent of the radioactive particles released from the accident were deposited within a 30 kilometer radius of the plant (about 19 miles). In general, therefore, the closer to the power plant, the greater the short and also long-term threat to physical well-being. However, the pattern of contamination within the 30 kilometer radius was very uneven, and some areas remained relatively clear. With 50,000 residents, the city of Prypyat was only 3 kilometers away (about 2 miles). Another 12,500 people in the town of Chernobyl were only 18 kilometers away (about 11 miles). And several hundred smaller towns and villages were located within about 60 kilometers of the plant (about 48 miles).

Arguably dangerous levels of radioactive contamination spread far beyond 30 kilometers. Moreover, a truly massive land area—one which included most of Europe as well as the European portion of the former Soviet Union—experienced measurable radioactive contamination as well. However, only the surface ground, water, vegetation, and buildings located within approximately a 30 kilometer boundary of the power plant became designated officially within the Soviet Union (May 3, 1986) as contaminated beyond acceptable levels. That area came to be referred to simply as "the zone." In the face of major controversies about thresholds of acceptable exposure, several countries in western Europe undertook actions such as destroying cattle, milk, vegetables, and other foodstuffs to avoid unnecessary doses to their residents.

Radiological monitoring and decontamination work by the Soviet Union began almost immediately. Within about a day or two, the duration of exposure of all workers involved in either damage control at the plant or decontamination activities in the surrounding area was being monitored and controlled to a certain extent. Some locations were so heavily irradiated that people could only work there for a few moments before exceeding what were defined as permissible levels of exposure. By 27 April, an emergency evacuation began for about 135,000 people living within the 30-kilometer zone. The mass exodus of so many people caused severe organizational problems (e.g. required document checks, reuniting separated family members, incessant inquiries from the outside) as convoys of people were moved to areas outside the zone. The World War II experience of massive evacuation was a cultural resource in this regard, and there is no reported evidence of panic.

According to official accounts of what happened, it took between eight days and two weeks to complete a successful evacuation. Official accounts conclude that the exposure of most individuals within the "zone" remained below a dosage level where medical intervention was required. According to unofficial accounts, it took well over a month to complete a rather tortured evacuation which exposed large numbers of people to unacceptable amounts of irradiation. And while there is considerable debate about how well managed the evacuation was, both official and unofficial accounts indicate that it was as much ad hoc as planned. The process began on 26 April in Prypyat, the largest population center (50,000 residents) closest to the plant.

It has been officially stated that all schools were ordered closed in Prypyat on 26 April, and residents told to remain indoors with windows and doors shut. There is much disagreement, however, over the way these directives were given, when residents received them, and how many people actually altered their everyday routines on that day. Several unofficial accounts indicate that the residents of Prypyat went about their daily business most of the day, seemingly unaware of the dangers of radioactive particles in their environment. Notwithstanding uncertainties about precisely what took place on 26 April in Prypyat, radiological monitoring data were indicating unambiguously that irradiation was rising beyond acceptable levels. Evacuation procedures (e.g. transport of evacuees, identification of safe routes, provision of medical teams, establishment of relocation centers) were therefore improvised during the night of 26 April and early morning hours of 27 April. The evacuation of the Prypyat began on 27 April. By 4 or 5 May, similar evacuation procedures were implemented for the town of Chernobyl and remaining population centers within the 30-kilometer zone. Sometime later, many villages outside this zone were evacuated as well. Administrative control of evacuation emanated from Moscow and Kiev.

Kiev itself, capital and major city of the Ukraine, and located only about 100 kilometers from Chernobyl (about 62 miles), was also affected by the accident. In this city of over two million people, there was a continuous washing of streets, entrances of buildings, clothing, and footwear during the month of May 1986. Only some of the children of Kiev were evacuated by whatever means during this same time period. These included children of Communist Party officials and those from families who were more familiar with the threat. And as the city itself continued to function normally, milk products were removed from markets, all food imports were checked, and children could be seen running around with sticks as if they were measuring radioactive contamination with a dosimeter. They were playing "radiation."

What was referred to officially as *the exclusion zone*, from 3 May on was an area with very few residents and little activity other than decontamination by early June of 1986. The city of Prypyat and the town of Chernobyl were virtually abandoned. All of their buildings, offices, and shops were locked and sealed. There were no domestic animals remaining. Both within and around these centers were large numbers of bulldozers, tractors, cranes, excavators, concrete carriers, and military freight carriers. Approaching Prypyat and Chernobyl were numerous dosimeter checkpoints, with very strict inspection of all vehicles. Soldiers wearing green antichemical protection suits decontaminated various pieces of equipment coming out of the zone.

As described above, the major activities during the immediate emergency period related to the makeshift smothering of the Unit 4 reactor, some initial radiological monitoring and clean-up work, and evacuation of population centers within and sometimes outside about a 30-kilometer zone surrounding the power plant. At least four major problems have had to be addressed since that time: first, the permanent containment of the Unit 4 reactor; second, the more systematic clean-up and decontamination of at least the 30-kilometer zone, and possibly an area extending as far as several hundred kilometers; third, restoration of housing for those evacuated; and fourth, determination of long-term medical consequences for people exposed to widely varying amounts of radioactive contamination. All of these tasks have had to be accomplished in the face of sustained scientific and news media scrutiny of an unprecedented event, followed by a major transformation of the entire society.

The long-term entombment of the Unit 4 reactor—which came to be called the Sarcophagus—was essentially an extension of the initial makeshift effort to prevent the further escape of radioactive materials into the atmosphere. The basic goal was to cover and seal the reactor forever, but there was no clear idea in the summer of 1986 about what the lifetime of entombment would be. Notwithstanding that uncertainty, a timetable for completion of a permanent concrete shell over the damaged reactor was geared to a schedule for returning the Chernobyl power plant back into normal service. The official deadline for both events became 1 October 1986, i.e. the Chernobyl Unit 1 reactor would not be restarted until the Sarcophagus was completed.

Constructing the Sarcophagus was highly complicated from both engineering and organizational standpoints. It required large amounts of equipment and, because the exposure of workers to radioactive particles had to be strictly limited, the sequential involvement of many thousands of people. The tomb was constructed of huge concrete blocks, each of which weighed up to several thousand tons. The blocks were assembled at some distance from the Unit 4 reactor. Once assembled, their transport to the reactor was extremely difficult because access roads were not designed to carry such heavy loads. On site, the blocks then had to be put into place with heavy-duty cranes. While all this was happening, radiological monitoring and decontamination procedures had to be maintained continuously.

By October, 340,000 cubic meters of concrete had been poured and placed, 3,000 tons of metal housing had been mounted, and the Sarcophagus had reached a height of 74 meters. While it is not clear when the sarcophagus was officially designated as completed, it is clear that the Chernobyl Unit 1 reactor was restarted on schedule in October. It is also clear that all work on the shell was finished by the end of December 1986. The last stage of construction involved the building of a 1-kilometer long fence around the Unit 4 reactor and the moving of uncontaminated soil to the area immediately surrounding the reactor. Since March 1987, a single diagnostic system composed of four computers has been housed in a new section of the power station and used continuously to monitor temperature and radiation levels in the entombed reactor. The entombment technology appears to have worked as planned, but how permanent it will be remains uncertain. Signs of extensive wear were being reported in the press by early 1991.

The massive clean-up effort focused initially on the 30-kilometer zone surrounding the plant. It involved three types of decontamination. First, anywhere between 10 and 50 centimeters (about 4–20 inches) of contaminated soil from the surface of between 400–1000 hectares (about 1,000–2,470 acres) was simply removed, placed into steel containers, and transported to about 800 special radioactive waste dumps. Because even the larger figure of 1,000 hectares is only a small portion of the land area in the zone, the cost of soil removal must have been an important constraint. The land actively decontaminated, therefore, was that closest to the power plant and to some population centers. Considerable uncertainty remains about the numbers and locations of the 800 waste dumps, and also how these obviously dangerous sites are being maintained over time. At least one primary waste dump was apparently located near the Chernobyl Unit 5 reactor, which was close to being completed at the time of the accident.

Second, water and special decontamination solvents were sprayed on buildings, roads, machinery, and equipment to remove radioactive particles. The contaminated water was then collected and buried to the extent possible. Third, the ground surface was sprayed with a solution which, when solidified into a film containing radioactive particles, could be removed and dumped at one of the waste sites. The majority of workers were engaged in this last type of decontamination. Major problems facing all decontamination efforts included lack of knowledge about changing conditions in the zone, the existence of hot spots of more intense irradiation, and the continuous churning up of dust by trucks and other heavy equipment.

As noted above, while the initial decontamination effort concentrated on the so-called exclusion zone surrounding the power plant, it was expanded from there as deemed necessary between 1987 and 1989, and this has apparently continued to this day. While the scale of the decontamination effort has been immense, with perhaps as many as 600,000 people involved in one way or another, less than 5 percent of the exclusion zone required serious counter-radiation measures (at a cost of more than one billion rubles). There is considerable debate about the extent to which decontamination has been successful. The debate revolves around the adequacy of decontamination activities relative to need. But because there is far from complete information about either, that debate is not likely to be resolved any time soon.

The evacuation of at least 135,000 people—some of whom moved more than once because their initial destination was unsafe or otherwise unacceptable—was obviously an enormous undertaking. It was decided to rehouse evacuees in regions immediately to the south of Chernobyl. In this way, those engaged in agriculture would remain in familiar territory, and logistical problems for all evacuees would be kept to a minimum. By September 1986, it was reported that a special government account of about 500 million rubles was being used to assist the victims of the Chernobyl disaster. It was also reported that most of this money was being used to finance the reconstruction of some 12,000 new houses. It appears that by October 1986 well over 7,000 new houses had already been constructed (between June and October) for evacuees from the 30-kilometer zone, and another 5,000 had been completed by the end of December that year. In addition, about 8,000 apartments had been made available by October 1986 in the cities of Kiev and Chernihiv. These units were allocated to former residents of Prypyat and Chernobyl. By April 1987, the number of apartments allocated had reached 13,500, and by the early summer of 1987 there were firm plans for an additional 3,000 homes and 1,500 apartments.

If the above figures are correct, then approximately 30,000 government financed housing units had been constructed, allocated, or planned for within a year of the accident. With an average family size of about three to four people, it would appear that the vast majority of evacuees were rehoused in an expeditious manner. Debates about the quality and administration of some housing settlements aside, most accounts of the housing program give it high marks. An obvious need was met in a timely and rapid manner. It must be noted, however, that some new settlements were located in areas that were contaminated (e.g. Slavutich, Ukraine, and a dozen settlements in Byelorussia). Ironically, in the same year that an ostensibly permanent rehousing program was being implemented, the Chernobyl units 1, 2, and 3 were restarted, and the possibility of reviving the entire 30-kilometer zone was being given serious consideration by Soviet authorities.

But that revival of the exclusion zone apparently had not taken place by the end of 1992 (when this case study was written). It appears that Prypyat had largely remained a ghost town. It was announced as early as July 1986 that a new town was going to be built on the border of the zone for workers at the Chernobyl power plant. It was supposed to become the residence site for about 10,000 people. But the project never got off the ground and the designated area for the town became only a dormitory settlement for about 2,000 workers. Then, in October 1986, plans were announced for the building of a larger replacement city to be located about 45 kilometers (about 28 miles) from Prypyat. Planning called for a thoroughly modern city of 30,000 people, one whose initial leadership would be elected by former Prypyat residents, and one which would have a museum at its center in honor of the heroes of Chernobyl. But problems of poor planning and design, inadequate administration, material and labor shortages, and insufficient services for those building the city plagued the project since the beginning. Its future character remains in doubt.

Recall that the smaller town of Chernobyl also was evacuated following the accident. It was then reported in December 1986 that about 3,500 people were working in shifts on clean-up within the town. By the spring of 1987, it was reported that background radiation had fallen dramatically in Chernobyl and that normal living conditions might be possible by the end of the year. It appears that the town had been divided into two zones by the summer of 1987 and the smaller of the two had been actively decontaminated; people were allowed to move about without respirators, even foreign journalists were invited to observe what was happening. Apparently, several thousand people have resided in this part of the city since 1987, most of whom work at the power plant. There is considerable debate about whether it has been safe for them to do so.

All things considered, most evacuees from the 30-kilometer exclusion zone who needed alternative housing were able to get it fairly quickly. The zone itself remained more or less abandoned, with access to it very strictly controlled. But there is one major center of activity within the zone which has continued with only a brief interruption since the accident—the production of electricity at the Chernobyl nuclear power plant. The commitment of the Soviet government to nuclear power remained very strong from 1986 until the nation was dissolved in 1991. But during that same period, internal concern about nuclear power grew a great deal, and this concern made a difference.

The head of the Soviet delegation to the International Atomic Energy Agency hearings on Chernobyl in 1986 died in April 1988. His memoirs, which were disseminated widely (published in *Pravda*), revealed a deep concern for the way in which nuclear power plants were being built and operated in the Soviet Union. By the fall of 1988 there was clear evidence that the country's nuclear program was being modified. A special commission headed by the vice-president of the Academy of Sciences concluded that the Crimean nuclear power plant should not be built because of its location in an area of high seismic activity. By 1989, plans for several new nuclear power plants in the Ukraine had been shelved or abandoned. Among them were the plans related to completing construction of the Chernobyl power plant (Units 5 and 6).

Finally, perhaps the most vexing problem of all is determining the long-term medical consequences of the accident. An important factor to remember is simply time. Most of the radionuclides released following the accident had a very short life wherein they could cause medical problems (a few days or weeks). But others can remain radiotoxic for a very long time (from many decades to centuries). While there is considerable knowledge about what radionuclides in varying doses can do to the human body, determining actual exposure rates in an affected population is a tremendous data-production problem.

It is relatively certain that about 50 tons of radioactive particles (about ten times the fallout from Hiroshima) were released following the accident. Some portion of these particles continues to be spread through the topsoil, vegetation, and reservoirs of a large area of the Soviet Union. Beyond medically established thresholds of exposure, such particles endanger the health of people who inhale or ingest them. However, patterns of contamination and exposure are so varied that it is virtually impossible to give a precise estimate of the number of people who are threatened, how far from the Chernobyl Unit 4 reactor that threat extends, and what the long-term medical consequences for these people will be.

Official Soviet estimates of these consequences have always been on the low side and loosely tied to the population evacuated, people directly involved in clean-up or decontamination work, or both categories because of their presumably higher exposure rates. For example, in an early (August 1986) report to the International Atomic Energy Commission, Soviet experts estimated among the approximately 135,000 evacuees, the spontaneous incidence of all cancers over the next 70 years is not likely to be increased by over 0.6 percent of what would have occurred anyway. The corresponding estimate for the remaining population in most regions of the European part of the Soviet Union was expected to be within a range of 0.03 to 0.15 percent. Estimates of health impairment due to genetic effects were even lower (20 to 40 percent of the estimated excess cancer deaths).

These and later official estimates have been challenged many times during the past several years, both inside and outside the Soviet Union. They are criticized as being exclusive rather than inclusive, and far too low for even the aggregates of people who were closest to the source of radioactive particles. However, the systematic evidence needed to resolve debates about long-term health effects is not likely to be available for quite some time, and given the major political transformation which has taken place in the former Soviet Union, perhaps such evidence will never be available.

CASE COMPARISONS

The two events fall readily within the above conception of disaster because they were sudden, involved unambiguous definitions of physical harm and social disruption at community and regional levels, and were catalysts for collective actions on a massive scale to meet immediate needs. The actions which occurred highlight vividly the adaptiveness of human beings and social structure during crisis conditions. There are many important differences in these events as well, and this is why life history and comparative sociological studies of disasters are so important. I will mention several of the most obvious differences for discussion purposes.

First, while certainly not everyday occurrences earthquakes have occurred repeatedly throughout recorded history. Nuclear technology is a twentieth-century phenomenon; power plant emergencies have been relatively rare since their construction began after World War II; and the specific accident at Chernobyl was unprecedented. Response to earthquakes is based, in part, on cultural advantages that have not evolved for nuclear accidents.

Second, definitions of physical harm, social disruption, and duration of impact take on new meanings when nuclear disasters are considered because of the presence of radioactive contamination. The threat of exposure to fission products for those working on site immediately after the Chernobyl accident was direct, invisible, and life threatening. The threat of exposure for anyone remaining within the 30-kilometer zone surrounding the plant was very dangerous as well, not detectable without special equipment, and longer term. It must also be remembered that the 30-kilometer exclusion zone was a somewhat arbitrary boundary. The threat of exposure to radioactive particles was arguably national and even international in scope.

Third, the rapid mobilization of human and material resources that is typical following natural disasters obviously was hindered by the presence of radioactive contamination. This is not to say that a truly massive clean-up effort was not implemented and sustained in this case—it certainly was. But the need to continuously monitor and control human exposure to radioactive particles meant that several hundred thousand workers were needed to complete activities that, in the absence of contamination, probably would have required only several hundred people. It remains debatable whether complete decontamination has been accomplished. Permanent entombment of the Unit 4 reactor is a continuing concern. Only partial, rather than total, decontamination of the 30-kilometer zone has been achieved. Relatively continuous radiological, as well as medical, monitoring is required.

Fourth, there are any number of historical precedents for large-scale evacuations such as the one following the accident at Chernobyl. They have happened during wars since time immemorial. They have also occurred routinely prior to natural disasters such as floods and hurricanes, or following accidents where toxic chemicals have been released into the atmosphere. The evacuation of the Chernobyl region was quite similar to evacuations generally—i.e. the process was orderly and there was little or no evidence of panic (Quarantelli 1954; Perry 1985). But what makes this disaster historically different is that the evacuation appears to have been permanent for the majority of those leaving the 30-kilometer zone. Most large-scale evacuations tend to be temporary, even during wars, with displaced people being able to return to their own communities. The situation here was very different.

Fifth, although the long-term recovery of Mexico City was never at issue, the city of Prypyat, the town of Chernobyl, and tens of smaller village centers had not recovered by the end of 1992, and their future remained uncertain. By taking a longer time-frame, it is certainly possible, of course, that the 30-kilometer zone surrounding the plant will become viable once again. But whatever happens in this zone, it is not likely to be described as a simple extension of pre-disaster evolutionary trends. This disaster event, at least, has been a catalyst for some very fundamental social changes. While these changes have been most apparent in the region directly affected by the accident, they do not end there. For example, prior to the dissolution of the Soviet Union, the expansion of its nuclear energy program had been reduced in a major way as a direct result of the accident. Now that the future of the new commonwealth itself is in doubt, one can only guess what will happen to nuclear power and the Chernobyl region.

Finally, there is the issue of disaster prevention. Although actions can and have been taken to mitigate their effects, earthquakes are interpreted easily by the public and political elites as "acts of God" that cannot be prevented. This interpretation seems compelling, at least up to a point. While some locations are more prone than others to natural disasters, the very act of living puts all human beings in harm's way to some extent. So when that unlikely event does occur, it is not surprising that victimization is perceived to be random. On the other hand, disaster prevention has always been a central concern with respect to peacetime uses of nuclear energy. Thus, to the question—Could the accident at Chernobyl have been prevented?—the answer has to be yes, lest the legitimacy of peacetime uses of nuclear energy be undermined completely. In this sense, the disaster at Chernobyl is similar to events such as airplane crashes, train wrecks, industrial accidents, and mine cave-ins, because the assignment of blame quickly became a central issue (Quarantelli and Drabek 1967).

CONCLUSION

One must conclude that the Mexico City earthquake and Chernobyl power plant accident involved notable similarities and important differences that merit attention by sociologists. Knowledge within this specialty can continue to expand by a comparative strategy that builds on the above conception of disaster as a catalytic event (Kreps 1984, 1989b). Because not all environmental hazards and other types of social problems have these definitional qualities, it is critically important to isolate those that do from those that do not. While being exclusive in this sense, the boundaries of disaster research remain very broad. They include any environmental, technological, or sociopolitical crisis that can be designated in social time and space.

Length of forewarning, magnitude and scope of impact, and duration of impact have physical and temporal dimensions that can be compared across whatever events that might be of interest. The social dimensions of disaster can be compared as alternative structural forms that occur before, during, and after these events (Kreps 1985). Of particular importance here is the use of basic sociological constructs (e.g. collective behavior, formal organizing, social networking, role enactment) to describe and interpret disasters as social constructions (Bosworth and Kreps 1986; Saunders and Kreps 1987; Kreps 1989a, 1991; Kreps and Bosworth 1993; Kreps and Bosworth et al., 1994). A central theme from previous disaster studies is that all crisis responses evidence patterns of structural continuity and change. The patterns that have been described can be explained in terms of variability of crisis events and social systems that create vulnerability to them (Hewitt 1983b).

Given the formidable research agendum facing those who study disasters, I hope that consensus about their fundamental nature can be maintained. Absent that consensus, this field will continue to have an identity problem within the social sciences. Certainly, that is not what the pioneers of this field intended.

NOTES

1 This section of the paper expands on my earlier article on the boundaries of disaster research (Kreps 1989b). That article was critiqued by Drabek (1989b), Turner (1989), and Quarantelli (1989a). The dialogue among us highlights several key conceptual issues facing disaster research.

2 This case description is extracted from Dynes, Quarantelli, and Wenger's (1990) detailed study of individual and organizational response to the earthquake. Their study is based on interviews with key organizational participants, a variety of documents, and two random sample surveys of the adult Mexico City population. The primary focus of their study was the immediate emergency (two-week) period following the earthquake. Survey data summarized by Dynes, Quarantelli, and Wenger suggest that official estimates of physical impacts

(deaths, injuries, property damages and losses) may have under-represented the extent of physical harm. I will report only the official estimates in the description which follows.

3 No brief description can capture the full complexity of this nuclear accident and the consequences which continue to flow from it. The literature on Chernobyl is quite large. I have developed a limited accounting of this disaster from the initial summary report of the International Atomic Energy Agency (1986) and subsequent books by Marples (1988), Shcherbak (1989), and Medvedev (1990). Each of these documents highlights continuing uncertainties about the physical impact and social disruption which immediately followed the accident, and the longer-term implications of what is characterized universally as the most severe peacetime nuclear disaster in history. I am in no position to resolve any of these uncertainties and conflicts. Instead, my goal is to communicate with as little distortion as possible a general sense of what happened so that this event can be compared sociologically with the Mexico City earthquake. I would like to thank Boris N. Porfiriev of the Russian Academy of Sciences for reviewing (and helping me to correct) an earlier draft of this section of the paper. Porfiriev was a member of a special investigation commission on Chernobyl created by the USSR Supreme Soviet in 1990. Any remaining technical errors are my own.

5

ISSUES IN THE DEFINITION AND DELINEATION OF DISASTERS AND DISASTER AREAS

Boris N. Porfiriev

We present an extensive discussion of the linguistic, conceptual and practical issues in discussing the category of disaster. It is argued that there are two principle orientations or approaches to research, namely, an applied/pragmatic one and a theoretical/conceptual one. These are based on ontological and epistemological grounds, respectively, which serve as the main factors determining the existing differences and variations in the studying and understanding of disasters. The other reason for the variation is the logical and terminological inaccuracy of individual researchers in reasoning on disaster matters. The categories of *ecological disaster* and of *ecological disaster zones* have been selected as critical cases illustrating that issue in the theoretical approach. Also presented is the concept of disaster area as a management objective, and the classification of territories based on the depth of the destructive impact on socio-ecological systems, and on the types and groups in such territories. Also, briefly discussed within the framework of the pragmatic approach are the principle measures and elements of an organizational system model for mitigating disaster aftermaths.

INTRODUCTION

In late 1992, the representatives of the Ministry for the Environment of the Russian Federation proposed that I analyze and find causes for the critique of its guide, *Assessment Criteria For Revealing Environmental Emergency Situations and Delineating Environmental Disaster Zones*. It had been prepared in early 1992 by some regional public administration and parliament officials responsible for environmental, emergency, and disaster management. Despite my recommendations, the voluminous document containing more than 300 pages, as shown in its publication in 1994 in *Zeleniy Mir* (Green World Weekly), remained practically untouched. It was thus doomed for further troubles in terms of its practical application, but also remained controversial. Scrutinizing the guide, questioning some of those officials who should use it, and, at the same time, studying the large amount of literature on the theory of disasters, revealed two important issues.

The first is one of the main reasons for the aforementioned troubles: the misleading and overcomplicated terminology, as well as the methods used for identifying disaster areas. The second is that, despite many years of intensive research, there still exists a wide gamut of interpretations of disaster that often look inconsistent and even contradictory. Therefore, both the basic question of "What is a disaster?" and the need for clarification of its conceptualization and definition repeatedly emphasized by a few respected scholars, remains an acute problem (see Kreps 1989b; Kroll-Smith and Couch 1991; Quarantelli 1985b). These two issues serve as a catalyst for my attempt to develop something that I wish to call a *system approach* to both disaster definition and disaster delineation.

DISASTER: LINGUISTIC, CONCEPTUAL AND PRACTICAL ISSUES

Early definitions of disaster were developed in explanatory dictionaries by linguists trying to marry folk and everyday usage of this term with somewhat canonical interpretations. Later on these definitions evolved together with changes in the world and language, but in essence they preserved the initial meaning of the subject. According to Webster's *New World Dictionary of the American Language*, a disaster is:

> any happening that causes great harm or damage; serious or sudden misfortune; calamity. Disaster implies great or sudden misfortune that results in loss of life, property, etc. or that is ruinous to an undertaking; calamity suggests a grave misfortune that brings deep distress or sorrow to an individual or to the people at large.

Whatever the classic definition of disaster, it varies substantially in the foreign dictionaries reflecting for each the particular country and the specifics of the culture and the language of each nation. According to the famous Dahl's *Tolkovi Slovar Zhivogo Velikorusskogo Yazika* (Explanatory Dictionary of the Great Russian Living Language):

> *beda* (misfortune) or *bedstviye* (disaster) is an incident, accident or other harmful occasion resulting in losses and sorrow. Disaster is more related to known cases of crop failure, epidemic, storms,

> floods. To suffer disaster means being in a disastrous or dangerous situation; badly needing something; dying.
>
> (Dahl 1989)

The modern *Dictionary of Russian Language* simplifies the term disaster by stating it is a big misfortune usually linked to natural hazards (Ozhegov 1987). It is interesting to note that Webster's definition of disaster treats it as a discrete event rather than a process or a state (condition) of affairs accentuating its physical impact, while the interpretation of calamity is biased towards its sociopsychological aftermath. So considering that the modern socially-oriented concepts of disaster are within Webster's linguistic framework, it seems more preferable to use the term "calamity" (e.g., see Hewitt 1983b) rather than "disaster." At the same time the Russian equivalent of disaster (*bedstviye*), while stressing the role of natural agents, as earliest definitions of disaster did, balances both physical and social aspects of disaster, paying the latter even greater attention, thus approaching modern interpretations of this term.

One should logically expect that irrespective of the circumstances, the core of these modern conceptualizations of a subject should be based on the aforementioned definition of disaster, or better, of calamity; that is, as an economically and socially harmful and ruinous happening or occasion. However, this is only partially true because researchers use different criteria, a methodology substantially predetermined by their professional background, and also because the phenomenon is the target of a particular study. Consequently, very often the elaborated definitions differ from one another, sometimes drastically, thus making difficult its precise characteristics in terms of the aforementioned encyclopedic designations (see Britton 1987; Drabek 1986; Kreps 1984; Quarantelli 1987a, 1987b).

To make a general comment rather than to state a reservation on this observation, I would like to note that in no way is it new or extraordinary for social science to operate with commonly used terms within a new research-specific context. As Engels put it in the introduction to Volume I of Marx's *Kapital*:

> There is one inconvenience we failed to remove for our reader. This is in using some terms with meanings different from both colloquial speech and those used in traditional political economy. But it is inevitable.
>
> (Marx 1961)

Claude Gilbert proposes to sum up all conceptualizations of disasters within three paradigms as follows. With respect to a disaster, it is: (a) the result of the impact of external forces; (b) the result of social vulnerability; or (c) the result of uncertainty.

Pelanda in turn suggested embracing all definitions of disaster into three categories basing on interpreting it as: (a) a result of negative social and environmental impacts; (b) a state (condition) of collective stress in a community; or (c) a contradiction between the capacity to cope with destructive agents and their negative impacts (Pelanda 1982a).

This typology reflects three conceptual dimensions: causal, descriptive and normative, which I believe to be absolutely correct, but somewhat insufficient and needing additional "axes."

In conceptualizing and defining disaster, some researchers prefer static rather than dynamic approaches and thus consider a disaster as a discrete happening. These researchers really base their definitions of this category on viewing it either as an *occasion* (Dynes 1988; Quarantelli 1987a, 1987b, 1992a, 1992b), or an *event* (Fritz 1961; Kreps 1989a, 1989b) or *phenomenon* (Horlick-Jones, Fortune and Peters 1991a, 1991b; Kroll-Smith and Couch 1990a). At the same time, a few of them also refer to the "principle of continuity", thus involving a timescale into their concept of disaster (Quarantelli and Dynes 1977).

Meanwhile, some scholars, using a dynamic approach to disaster, treat it as a category of *action* (Dombrowsky 1981), while others try to combine both of the aforementioned approaches within some kind of generic or ecological concept that considers disaster as a certain *social state/condition*; i.e., crisis or vulnerability (see Barton 1969; Gilbert 1991; Kroll-Smith and Couch 1991; Lagadec 1991; Pelanda 1982b).

Further, developing from this "nucleus" to an extensive definition of disaster, and starting from the earlier mentioned causal logic, some of them, primarily geographers, consider disaster as a *reason* for the following social disruptions or disorders (Alexander 1993; Kroll-Smith and Couch 1990a; Foster 1990). This description sets forth an exogenous or an outward type concept of disaster that designates it by physical agents that usually come from nature (i.e., outside the community), and thus treat it as a natural disaster similar to the older Dahl's dictionary version or everyday conversations.

Social scientists using the same causal logic display a wider gamut of interpretations of disaster. They consider it as a *social construction*, though all of them fit their view within an endogenous or inward-type conception of disaster. A few of them treat disaster as being the *result* of social processes or *social consequences* that create (give birth to sources of danger or hazard), or increases the vulnerability of a social system to hazardous impacts (Pelanda 1982a; Perrow 1984; Tierney 1989; Quarantelli 1987b, 1992b). This vulnerability may display itself either in the form of a collective stress situation (Barton 1969) or crisis within a social system per se (Lagadec 1988; Rosenthal 1986), or inconsistency between the capacities of the latter and the demands of a disaster situation in terms of both response and recovery (Kreps 1984; Pelanda 1982b; Turner 1978, 1979).

Other scholars consider disaster as involving a *combination of reasons*, whatever their origin (natural, technological, social, political) and *results*, both physical and social (see Horlick-Jones, Fortune, and Peters 1991a, 1991b; Kreps 1989a, 1989b). The latter approach for a long time has also been supported by an outstanding American geographer, Gilbert White. It should also be added that, along with the descriptive and normative interpretations of disaster embedded in Pelanda's typology, it is sometimes viewed as a *statistical category*. For example, UNDRO designates a disaster as:

> the probability of occurrence within a specified period of time and within a given area of a potentially damaging phenomenon (1982).

But Dynes (1993), among four different meanings, stresses three as indicators either of physical or social damage or negative evaluation in general.

Maybe somebody would see this in a different way, but I do not consider the aforementioned variety of conceptualizations and definitions of disaster as absolutely and really contradictory and inconsistent. As already mentioned in the introduction to this paper, they *look like*, they are contradictory. Though they are really different and thus often noncomplanary, there are still substantial grounds to treat them as complementary, i.e., lying within the same categorical domain (to continue a mathematical analogy). As Quarantelli and Kreps put it, there is considerable agreement or much more agreement than disagreement among scholars about the essence of what a disaster is (Quarantelli 1992b; Kreps 1989b).

With minor exceptions, I believe that the earlier speculations in this paper are evidence that all "deviations" from encyclopedic explanation of disaster reflect the professional backgrounds of the researchers, the specific tasks or goals of concrete studies, as well as the indistinct terminology of individuals, rather than in a discrepancy in the very meaning of disaster. Some scholars starting to answer to the question of "what a disaster is," further simply substitute or confuse it with other questions such as "what a disaster does," or "how society acts under conditions of disaster" that should of course be considered as relevant, but still definitely not addressing the original question.

That is why I argue that while further discussion on conceptualization and definition of disaster is underway, one of the most important issues to be considered is a very clear delineation of initial or starting premises, and the limitations of each study. First of all, it is crucial to distinguish the task function or, in other words, *the applied/pragmatic and theoretical/conceptual orientations* based respectively on ontological and epistemological grounds.

Within the latter category, a social scientist or a system analyst tries to develop a conceptually-based definition of disaster as such, using either broad (system or interdisciplinary) or specific (disciplinary) approaches, and attempts to balance cognitive, including psychosociological and empirical aspects. Within the pragmatic orientation of a study, its author who is normally a decision-making analyst (supporter), engineer or natural scientist (geographer, geologist, etc.) attempts to develop a rather operational framework or/and definition of a disaster necessary for elaborating laws, rules, regulations, and also for plans and measures to prepare, respond to, recover from and mitigate concrete disasters (earthquakes, floods, explosions at nuclear power plants, etc.).

The *research methodology and methods* of the two different types of studies on conceptualizing and defining disaster should also be expected to be quite different. The conceptual-focused ones would involve generic-type complex approaches and theories (system, sociological, linguistic, information, risk, multi-criteria decision-making in conditions of uncertainty) reflecting both objective and subjective (perceptual) aspects of disaster. At the same time, a pragmatic-oriented study would rather apply common sense, as well as widely known intuitive and simple logic approaches based on a limited number of objective criteria. These are aimed at explicitly revealing those basic features and stages of life history of specific disaster that permit developing the aforementioned normative and regulatory documents.

No less different would be the *form (shape) of the outcome or results* of the mentioned functions or orientations of the study. In a pragmatic type study, it may vary from a "working" definition of disaster simply to introduce further the action program or measures proposed (see the UNDRO definition of disaster above), to presenting a framework of the noted documents; from narrow or subjective versions of what a disaster is, sometimes with explicit quantitative criteria (see Clement 1989), to recommendations for and concrete drafts of laws, rules, regulations, or plans. The criteria may include: (a) types of hazards (disasters) sources; (b) the time for preparing; (c) the duration time of the emergency; (d) the magnitude of impact both in terms of territory jeopardized, population and ecosystems affected, damages and losses inflicted, and the political, psychological, technological and economic aftermaths; and (e) the time and resources needed for recovery. In other words, socioecological (including mortality and morbidity aspects), sociopsychological, socioeconomic, sociopolitical and organizational criteria may be used (Porfiriev 1989, 1991).

As to a conceptual-oriented study, one would get an extensive definition of disaster, distinguishing it from both routine and other abnormal occasions (emergencies, calamities, catastrophes) based on a verbal description with all or a few qualitative (implicit) criteria mentioned above. Within this kind of study, I would like to introduce my version of a definition of disaster, and to use a system/descriptive approach. This treats disaster as:

> a state/condition destabilizing the social system that manifests itself in a malfunctioning or disruption of connections and communications between its elements or social units (communities, social groups, and individuals); partial or total destruction/demolition; physical and

> psychological overloads suffered by some of these elements; thus, making it necessary to take extraordinary or emergency countermeasures to reestablish stability.

This definition, though relatively extensive, does not of course reflect all the variety and complexity of existing and potential disastrous states/conditions. However, from a methodological viewpoint it is important in that it seems to allow a delineation between routine and stable situations and disaster, on the one hand, and, on the other hand, between disaster and other forms of extraordinary situations irrespective of their causes.

The disastrous state/condition described, or a disaster, may arise from natural, technological, and sociopolitical circumstances, or a combination of these external (exogenous) and internal (endogenous) factors that make a social system vulnerable to losing its stability. I believe that the latter category, *stability*, well known as a crucial characteristic of a system reflecting the dynamic balance of its elements and linkages between them, should be considered as a basic principle for defining disaster and pointing out its main distinctive features. These include: (a) a breaking of the collective routines of social units (i.e., communities and societies), and (b) an urging of extraordinary, often unplanned measures for the survival of the latter and managing (reestablishing) the situation.

The (a) descriptor distinguishes disaster from emergency. The latter classically is defined as an unexpected occurrence or sudden situation that requires immediate action (Dahl 1989). It may involve communities (as a disaster does) or individuals (which a disaster does not). Besides, unlike an emergency, I believe unexpectedness not to be a necessary attribute of a disaster (e.g., in case of a famine, or a disaster caused by continuous toxic contamination), though sometimes it may be present.

The (b) descriptor separates a disaster from a crisis or a crisis situation. The latter is normally treated as a hard and complicated situation (Dahl 1989), or a turning point, a decisive, crucial time/event, or a time of great danger or trouble with the possibilities of both good and bad outcomes. Our historic experience of economic crises, for example, shows that such a situation may last for months, maybe even years, and unlike a disaster or an emergency does not require urgent (within 24–48 hours) or short-term (within a few days) measures to save people. Rather, the emphasis is on midterm actions to reverse the situation to the better (Porfiriev 1991).

The aforesaid does not mean that I am absolutely neglecting the category of *vulnerability* which is emphasized by many scholars. I also consider it a very important idea for the conceptualization of disaster. The idea of vulnerability is a theory or theoretical approach to explaining the origin (causes) and development of disaster, and thus answering the question of "why a disaster" rather than "what disaster is" and "what disaster does," that are often the principles used in the definition of disaster. As Quarantelli put it:

"We should stop confusing antecedent conditions and subsequent consequences with the characteristics of a disaster" (1987b: 7). I guess the term "stability" would be quite adequate in this respect, in particular, considering it as the organic criterion of a social system, which is the main target of a disaster.

Whether apt or infelicitous, the proposed definition of disaster has been elaborated within a conceptual-oriented approach, and consequently should in the best way be used for that type of research. The next section of the paper deals with what has been earlier labeled as a pragmatic-type study.

DEFINITION AND DELINEATION OF DISASTER AREAS: METHODOLOGICAL APPROACHES

The defining and declaring of a disaster area—which represents a rather complicated research issue that requires both an analytical and a system approach—is at the same time a very practical problem. This directly stems from the needs of the legal and executive branches of local, regional or federal public administrative bodies who undertake activities targeted toward setting up the boundaries and status of the disaster area; that is, the territories suffering from natural or technological disasters. The issues of evacuation, sheltering, the providing of medical and nutritional services, the providing of pensions (or loss compensation), and the state of the environment are defined in locational terms, and serve as the focus of those officials responsible for making decisions.

In establishing disaster areas, the origin of the causes of the disaster is of no great importance. What is really paramount, on the one hand, is the social disorder, the disruption of the routine life of communities including the presence of dead bodies, the increased morbidity, the material damages and losses, and the psychological stress inflicted by certain destructive forces. On the other hand, there is the concern resulting from the community's own vulnerability caused either by contradictions within the social structure, or by loopholes in the planning and preparedness activities.

As far as the natural environment is concerned, in contrast to communities, it is practically always jeopardized by the potential impact of physical agents both on inhabited or uninhabited territories. From this viewpoint, any "disaster" or emergency should be considered as ecological because others simply do not exist. If anyone wants to discuss the sources or risks coming from the natural environment, rather than the objects (communities or ecosystems) involved or the impacts produced on them, and to stress that the destructive agents (or disaster in its traditional meaning) come from the ambiance, then it is quite logical and far preferable and clearer to use the well-known term of "natural disaster" rather than misleading categories of *ecological disaster* and/or *ecological emergency areas*.

Considering the aforementioned, the categories that are extensively used in the Environmental Protection Act of the Russian Federation (EPAR) enforced since 1991 should be qualified as being both artificially and methodologically insolvent. They come into contradiction with the rules of creating categorical systems or frameworks which prescribe that each innovative term introduced requires its distinctive antonym, which in this specific case is lacking. As a result, discrepancies and misunderstandings frequently occur both in theory and practice, in normative documents in particular, including the EPAR. To avoid all these problems and headaches of decision-makers, I believe it would be reasonable to keep applying the widely accepted categories of "disaster(s)" and "disaster area(s)". However, the latter should be clearly and strictly defined using specific criteria and parameters allowing a delineation of these areas.

I suppose that the degree of social disorder and environmental impact present should serve as a crucial, integral, qualitative criterion that is also most convenient for decision-makers (public administrators, disaster managers, and so on). Using it together with an indication of the categories of destructive agents is a good basis for developing a rough typology of the territories and approximating the definition of and classes of disaster areas (see Table 5.1).

Table 5.1 Typology of affected areas by health consequences

Categories of the territories	*Healthy*	*Unhealthy*	
		Crisis Areas	*Disaster Areas*
Destructive agents			
Natural including:			
permanent impact	–	[+]	[++]
temporary impact	–	+	++
Anthropogenic (technological and sociopolitical):			
permanent impact	–	[+]	[++]
temporary impact	–	+	++
Combined (synergetic) including:			
permanent impact	–	[+]	[++]
temporary impact	–	+	++

Notes
– Nil or negligible social disturbances and/or environmental impact
\+ Notable/substantial social disorder and/or environmental impact (close to threshold loads and acceptable risk levels)
++ Considerable/great social disorder and/or environmental impact (tolerable loads and acceptable risk levels exceeded occasionally for many times

Following this classification, all territories may be subdivided into two main *types*:

1 *Healthy areas*, where communities and ecosystems are not or only slightly jeopardized by natural, anthropogenic (technological and sociopolitical) or combined hazardous impacts, and consequently social and environmental disturbances are practically nil or negligible. Those areas constitute only a small percent of the total territory of Russia at best.
2 *Unhealthy areas*, where the situation is just the opposite, and that embrace the bulk of the territory of Russia. This should be further subdivided into two classes: *crisis areas* or areas of crisis situations; *disaster areas* that are distinguished according to the degree of gravity of the impact on communities and the environment.

If the probability of human death is taken as a criterion for the destructive or destabilizing impact as the aforementioned *Criteria* guide proposes, then the three classes of the territories that I suggested earlier (wealthy, crisis, and disaster zones or areas), would include five types of regions. These are delineated in the guide, creating together the class of disaster zones (including all elements in Table 5.1 marked ++).

At the same time, specifying the impact of "pure" environmental factors on mortality and morbidity should be considered as very difficult (if at all possible) while delineating disaster areas in practice (especially considering the urgency and extreme conditions for a decision-maker). Besides and along with environmental factors, the other mortality and morbidity causes (physiological, social, psychological) are only slightly or completely not connected with the state of the environment act. That is why I believe it to be artificial both to consider the guide's criterion (the mortality indicator) as environmental risk, and to delineate the categories of *ecological disaster* and *ecological emergency areas* mentioned earlier. Additionally, I consider it incorrect to restrain the environmental risk by mortality and morbidity aspects, leaving aside the impact on ecosystems.

Summing up the aforesaid, the disaster areas may be treated as territories impacted by a disaster (for the definition of disaster, see the earlier discussion). They may be subdivided into two groups as depicted in Table 5.2:

1 *Emergency or acute disaster areas* induced by natural, technological, social (sociopolitical) or combined temporary (immediate or periodical) action, factors or impacts, including: floods, earthquakes, tornadoes, winter storms, explosions, wildfire, etc. These areas are marked with + or ++ under "Unhealthy Disaster Areas" in Table 5.1.
2 *Chronic disaster areas* that could also be called areas of long-term disasters result from the permanent or prolonged action of the aforementioned

Table 5.2 Characteristics of declared disaster areas

Categories	*Types*	*Territory covered*	*Level of authority*
I	Local	District or their clusters within a town, city, SMSA, county (rayon)	Local authority (mayor, prefect, etc.)
II	Regional	A cluster of a few counties *(rayons)* state(s) *(oblast krai)*	Regional authority (state governor; *oblast krai* administrator)
		Independent (sovereign) territories, autonomous and other republics within the Russian Federation	President(s), heads of the governments of the autonomous and other republics and independent territories within the Russian Federation
III	Federal	The same or the cluster of the aforementioned republics and territories	Federal President or Prime Minister

factors or impacts including, for example, desertification, famine, radioactive contamination of the environment, etc. These areas are marked with [+] or [++] in the columns under "Unhealthy Disaster Areas" in Table 5.1.

An analogous typology is presented for crisis areas. They can be subdivided into acute crisis and chronic crisis zones. In Table 5.1 they are marked respectively (with + and ++ and [+] and [++]) under "Unhealthy Crisis Areas".

Emergency and chronic disaster areas, while representing different groups within the same class of the territories, should also be treated as two phases or stages of the social, as well as economic, disorder and environmental degradation processes on a certain territory. The Chernobyl disaster can serve as a good illustration of the point. Within the first months after the reactor explosion, the radioactive fallout created a disaster situation in the nearest (30-km radius), and later in the more distant areas, that all presented grounds to be treated as emergency or acute disaster zones. In several months, and in some territories, even after some years, following the evacuation of the residents of the affected communities, deactivation of soils, etc., those territories were transformed into chronic disaster areas.

Consequently, emergency (acute disaster) and chronic disaster areas are never synchronous, that is, do not emerge simultaneously, though they often are centipoise or coincidental—that is, they occur at the same place. The degree of social as well as economic disorder and environmental degradation generally expressed through the total or aggregate risk (R) indicator (both for communities and the environment) is an important initial and necessary, but insufficient, criterion for delineating a certain group or category of disaster areas. In using it for operative administrative and executive measures, the decision-maker then should involve a number of various and more sophisticated indicators to specify the status and borders of those areas.

The issue of these indicators requires special analysis, and will be discussed further on in this paper, but it seems reasonable to make a short comment at this point. First, it seems necessary to decompose the very criterion (indicator) of total risk for communities and the environment into several qualitative and quantitative parameters or descriptors. In particular, these parameters represent medical (somatic and genetic), sociodemographic, and sociopsychological effects on individuals and communities. Among those parameters, the impact time or span of the exposure of humans and the environment to the hazardous and toxic factors, reveal both the scale and degree of the R indicator.

Second, the latter factors should be supplemented by a number of other criteria: the scale or share of the envelopment of a territory and community by a disaster, reflecting the magnitude or intensiveness of its destructive impact, and thus the disaster's category (possibly analogous to the Richter earthquake or other scales).

Social disorder, as well as economic and sometimes political disturbances, along with environmental degradation typical for a disaster, suggest a cluster of measures targeted for mitigating the situation. This does not always mean total elimination of the aftermath both for communities and the environment. Disaster areas were caused by the 1957 Chliabinsk (Kyshtym) and the 1986 Chernobyl radioactive explosions. These resulted in a fallout of long-lasting radionuclei and the subsequent emergence of new geochemical provinces, which created grave and untreatable sickness, not to say dead bodies, which serve as a vivid illustration of the point.

The aforementioned cluster of mitigating measures embraces two basic groups:

1 *urgent (extraordinary) short-term actions* that are usually linked with emergencies and emergency (acute disaster) areas. However, in cases of the permanent impact of disaster agents or destructive factors, when a disaster almost immediately becomes chronic (e.g., the Aral Sea region), these measures should also refer to chronic disaster areas.
2 *medium-and long-term actions* that are normally connected with the latter areas, and frequently occurring at the termination (final) stage of the emergency situation, if total elimination/liquidation of its aftermath is impossible.

Declaring the impacted community or region a disaster area should be considered as one of the first steps among the urgent measures undertaken. Existing experience shows that the executive branches of public administration (the authorities) are legally responsible for that action. Depending on the magnitude of a disaster, the means and forces to cope with it, and the degree of responsibility of the authorities, disaster areas may be declared at local, regional, and federal/national levels.

Local and regional levels of public administration, which bear the bulk of decision-making responsibility and performance in disaster areas and where the decisions are developed and put into action by local authorities (and their special/emergency services), are themselves determined by the concrete situation and current laws and regulations. The latter, in the leading countries of the world, are substantially developed and consider practically all types of disasters.

In case of a natural disaster, a state governor or a city mayor in the United States would act on the basis of 1936 laws, the 1972 flood management regulations, and amendments in 1977 further regulating earthquake hazard reduction. Unfortunately, up to now, Russia lacks these and some other paramount normative acts, which make local and regional authorities apply less concrete or specified acts such as the 1992 Security Act of the Russian Federation, and the 1991 Act on State of Emergency of the Russian Federation (in terms of disaster area regulation and management).

In the declaring of disaster areas of I and II categories (see Table 5.2), local and regional officials issue warnings both for communities and higher authorities, including the federal government and the president. Usually, when a serious accident occurs at a hazardous facility, it is necessary first to declare an emergency, and then a disaster, and to delineate disaster area laws and regulations and to put responsibility for warnings, the updating of information, and the evacuation of the operating organizational personnel on the owner of the facility. The latter are also responsible for warning local authorities, the mitigation or liquidation of the aftermath of the accident directly at the facility according to the operation (emergency) plan that should be necessarily developed by the owner or user of the utility, and then putting it into action together with the fire, medical and other special services of the city, or if needed, at the prefect's or mayor's request, of the region (county, state).

But there are *local authorities* who are legally responsible for official declaration of a disaster or disaster area status at the community (local) level. In Russia in particular they should:

- warn local citizens about the danger/hazard that has occurred and inform in a timely way regional and federal authorities and their emergency management branches;
- remind citizens about individual means of self protection and evacuation before the arrival of the rescue services;
- establish operation headquarters that coordinate the collective protective measures, including evacuation and rescue operations, together with

emergency plans developed by the staff of the authorities, primarily by civil defense and special services specialists;

- declare the Category I status, delineate the borders of the disaster area, and coordinate recovery activities within this area.

If the scale and magnitude of a disaster exceeds the capacity of the local authorities to cope with it, at their request the *regional* (*oblast*, *krai*, republic) *authorities* come to the scene. The head of the regional administration declares a disaster status on a certain territory and delineates (changes the status, broadens) the Category II disaster area, in particular by issuing new, higher status for the territory earlier classified by the local authorities as a Category I disaster area (see Table 5.2). As at the local level, an operation center (headquarters) which coordinates special services, voluntary/emergent groups, and the activities of individuals should be created. This work should be performed according to emergency plans developed by the aforementioned specialists of the regional administration.

In cases when the regional administration has insufficient capacity to cope with a disaster and/or manage the Category II disaster area, it may follow two options:

1. Appeal to the federal authorities (the government and the president) to declare a federal (national) disaster, and the respective territory a Category III disaster area. As world experience has shown, that appeal should be directed first of all to the officials and sections of presidential and governmental offices responsible for security issues. In Russia, that would be the President's Security Council, including—following the 1992 Security Act of the Russian Federation—the heads of the Department of Defense, the Department of Civil Defense, Emergencies and Natural Disaster Response; the Departments of Health, Environmental Protection, and some others. The Security Council arranges and undertakes intensive consultations, and formulates recommendations for the President, who is responsible for making the final decision on declaring the impacted territory a Category III or federal disaster area, and thus providing necessary emergency material and technical support from the federal reserves.
2. Address the matter of providing assistance to the authorities of the neighboring regions, guaranteeing analogous support to the latter in case of disaster. To have legal grounds for that process, cooperative or mutual assistance agreements between cities, regions (states, oblasts, etc.) are made beforehand. For example, such formal agreements are widely known and used in the United States embracing 48 of the existing 50 states, as well as a majority of counties and municipalities of that country. In Russia, though, these are only emerging and thus disaster management has so far been kept far more centralized. These

agreements are of special importance in case the president gives a negative response to the aforementioned appeal and, as the existing experience demonstrates, it is not quite a rare occasion. Continuing the earlier example, the President of the United States rejects up to 40 percent of the appeals of the state governors to declare a national disaster or disaster area in a certain state.

After the urgent measures are fulfilled and depending on their efficiency, two possible scenarios may take place at a disaster area:

1 effective recovery works, leading to a deferring of the disaster area status upon the decision of local/regional authorities in the case of Categories I and II territories, and of the federal president for Category III areas;
2 continuation of the disastrous situation for both communities and the environment over a long period, which stimulates a change of the existing status of the impacted territory to its declaration as a chronic disaster area (i.e., transferring it from a Category I to a Category II area) by the appropriate level of authority. The latter also develop mid- and long-term rehabilitation/recovery programs that prescribe successive mitigation measures and the elimination of the disaster effects. The status of a chronic disaster area which determines the utility regime for an impacted territory, the social protection measures, its environmental rehabilitation, or vice versa, the closing (classifying) of the territory and the permanent evacuation or moving of people from the disaster area, should be normatively established.

Section II of the 1991 Act of the Russian Federation on Social Protection of Citizens Who Have Suffered from the Chernobyl Disaster's Radioactive Fallout, may serve as a good illustration of the point. It established the following categories of respective territories or zones:

- closed (permanently abandoned; for decades, maybe for centuries);
- resettling/resettled;
- permission for living with a right to abandon for a certain monetary compensation;
- having privileged socioeconomic status, provided for by special donations from the federal budget.

Following Table 5.1, the last two categories should be denoted as crisis areas, and the first two as disaster areas. In the latter, in radiation disaster situations, living is prohibited and permanent residents (inhabitants) should be definitely evacuated (resettled) with economic activities strongly restricted in closed or permanently abandoned areas contaminated by Cs-137 over 40 Ci/sq. km or by Sr-90 over 3 Ci/sq. km or by Pt-239/240 over 1 Ci/sq. km inducing individual irradiation doses exceeding 5 mSv per year.

In the areas with a cesium contamination level between 15–40 Ci/sq. km or respective densities of strontium and plutonium (relocated area), living is permitted only in case of obligatory and regular radiobiological and medical testing and treatment; rigid protection measures to decrease irradiation loads are provided by local and regional authorities. Communities, as well as individuals, are informed through the mass media about those measures and activities in a disaster area. The citizens right-to-know about what is going on and being done in the territories affected by the fallout from Chernobyl are also guaranteed in the previously mentioned 1991 Act on Social Security, Sections 11 and 46.

CONCLUSION

The analysis of existing methodological approaches for developing definitions and concepts of disaster, its definitions and conceptualizations per se, as well as their applicability for practice (decision-making), reveals that most of them rest on certain theoretical and empirical grounds reflecting various values (professional, cultural, functional) and other orientations and different aspects of this complex objective. I believe that while some of those definitions and concepts have common features in general, they have been developed and formulated within different domains ("systems of coordinates") reflecting the professional backgrounds of researchers, the specifics of concrete study task goals, indistinct terminology, etc.

So, according to the relativist theory, at least some of them cannot be congruent; that is, what should be considered correct within one framework is not right within the other. Stemming from this point, I argue that any disaster research should start from a clear-cut delineation of this system, a formulation of the orientation of the study, its targets, limits, methods of analysis, while making necessary reservations and comments. That will contribute to both clarifying the systems of reasoning and arguments, enforcing their convincing effect, and finding bridges between those different systems, thus allowing accurate transfer from one conceptualization or definition to another.

These are not academic speculations or a play on words, but rather a demand from the theory and practice of decision-making, first of all, from legal (law-making) activities. The latter surely need very distinct, noncontradictory, single-meaning and juridically-correct formulations to provide for firm grounds for the enforcement of laws and the control of their implementation. The cases of the 1991 Environmental Protection Act of the Russian Federation, and the Assessment Criteria for Revealing Ecological Emergencies and Delineating Ecological Disaster Areas, clearly show unclear double-meaning definitions of the key categories, and affect in a negative way their effective application. That is why I feel myself quite critical toward a few parts of these documents, in particular to the sections referring to ecological disasters and areas of ecological disasters.

At the same time, I am far from considering exhaustive the proposed original definition and conceptualization of disaster, based on stability as a key category. These are rather sketches than fully-shaped constructions. Nevertheless, I believe them supporting, in terms of serving somewhat as a "shell" or bridge for the existing variants of definitions and theories of disaster, and thus worth further development within the conceptually-oriented research or approach. On the other hand, from an applied or pragmatic viewpoint of disaster management, the development of disaster conceptualizations, including disaster typologies, should follow the path of and refer to the criteria of being operational and specific, rather than exhaustive or universal. These should give priority to simplicity of understanding and decision-making, embracing management organizational structure or levels of making and implementing decisions, as well as the cost-effectiveness of the realization of the chosen strategy (policy).

FIRST REACTION ARTICLE

6

EXCLUDED PERSPECTIVES IN THE SOCIAL CONSTRUCTION OF DISASTER

Kenneth Hewitt

> [O]n closer inspection, it turned out that we also need to take into account the problematic of government. In short, one needed to analyze the series; security, population, government.
>
> (Foucault 1991: 87)

The purpose of this paper is to review and respond to the preceding articles that appeared in the special issue of the *International Journal of Mass Emergencies and Disasters* (IJMED). My principal charge is to look at the authors' approaches to answering the question of "What is a disaster?" and respond to their considerations. In doing this, I have outlined a variety of what I believe to be *excluded* perspectives in these formulations of what constitutes disasters.

INTRODUCTION: SOCIAL SCIENCE PERSPECTIVES ON DISASTER

My task in these reflections on the question "What is disaster?" is to respond after reading the other five papers. Answers could take several forms. They may be the purely definitional, seeking to establish and encompass the object of interest; or to list and classify its content. Alternatively, the focus may be how to explain or interpret calamities. It may be about social responsibilities and what Dombrowsky (1995a: 241) identifies as *programmatic declarations*. These days, the desire to establish professional "turf," can take precedence, trying to show disaster work and agencies having special roles and territory.

Something of all those considerations is found in the papers here. But their preoccupation is with social perspectives. Among the seven "ideal type terms" Quarantelli (1982: 458) once used to identify the scope of disaster work, at most three are considered: in brief, social disruption, the social construction of reality and political definitions. Moreover, the authors are less concerned with simply defining "disaster"—though Porfiriev makes a valiant attempt to resolve that one—than with Quarantelli's "question behind the question" (ibid.: 456). That is, with the preoccupations and paradigms that shape how we approach disaster. In this regard, the papers seemed to me to be preoccupied with problems arising from the relations between (Western) governmental and professional visions of the field. In that context, two of the authors look for greater refinement of existing practices, three seem to find existing approaches inadequate and propose or anticipate major changes.

My own background is research into the physical environment and geohazards. Yet, I have come to believe that social understanding, and socially just and appropriate action, are the more crucial issues for the contemporary disaster scene. And while I would welcome a clearer picture of the social content of this work, like the other authors, the main problematic for us does seem to be the social construction of disaster. That concerns the often covert or taken-for-granted way in which social conditions or "realities," shape how we think about and act toward disasters. In the fashionable terms of recent debate, it is to examine the relations between discourse, ideology and practice.

It should be said, at once, that "disaster" has taken its place among a number of not unrelated notions, such as the environment, development, population, poverty, security. They are commonly discussed as if referring to self-evident and age-old empirical realities. Yet, they have been radically redefined to suit modern instrumental agendas, linking science to professional and administrative practice. If not exactly coherent fields, each has become the focus of a community of researchers and publications, agencies and practitioners. As Gilbert stresses, it is since the Second World War that these notions have taken on their new meanings, and been deployed as one commentator puts it, according to "how the rich nations feel" (Sachs 1990: 26). That is the context, I think, that explains why the social constructions in these essays address, or revolve around Gilbert's uncompromising remark that:

> [The] scientific approach to disaster is therefore a reflection of the nature of the "market" in which disaster research became an institutional demand.
>
> (1995: 232–233)

For my part, I am less at odds with what the other authors say on that, than with some missing ingredients and perspectives. In some cases, that is because these authors have already made a major series of contributions which they do not wish to repeat here. Nevertheless, I think that some of those absent issues and literatures should be raised, and others suggest questions that will, hopefully, stimulate a discussion.

SOCIAL PROBLEMS WITHOUT SOCIAL CONTENT?

A broad range of studies in the 1970s and 80s seemed to demonstrate the overwhelming significance of societal conditions in the incidence and distribution of damage in disasters. My own field work on various of types of natural disaster, and disasters of war in a wide range of settings, reinforced that. It showed that whether, where, how and, especially, to whom disaster occurs, depends most closely upon established social conditions and controls over the varying quality of material life. The distribution of human casualties related especially to socioeconomic status. Material losses were often disproportionately concentrated according to age, gender, occupation, social position and, above all, lack of wealth and political voice. Even the effectiveness of risk assessments, warnings and emergency preparedness, depended most on whether or how they are (least) available to those most in need of them. Yet, a sense that these problems are primarily societal hardly describes the main focus or agendas in disaster work. Rather, there is a continuing dominance of geophysical, technological, and formal organizational models. It seems to me we have to address that to understand the contemporary social construction of disaster.

The physical agent or hazards paradigm

None of the other authors subscribes to a view of disaster in which physical agents define the problem. Some question or distance themselves from what Quarantelli has called the *agent-specific approach*, or taxonomies of disaster based upon it. Gilbert suggests that "the new paradigm" of disaster rejects the primacy of external agents as the source of disaster, and defense against them as its solution. Rather, the origins of disaster are seen in upsetting of human relations and social vulnerability. Kreps stresses the importance of disasters as social catalysts, in keeping with his earlier and more radical assertion that "disasters both reveal elemental processes of the social order and are explained by them" (1984: 309). Dombrowsky (1995a: 244) finds the key in an improved "organizational sociology," and Horlick-Jones (1995: 311–312) goes as far as to see "modern" disaster as the arena of "existential trust" in or "betrayal" by public institutions. As statements about the problematic nature of disaster from a social science perspective, I find these justified and very timely concerns.

However, it is a mistake to blithely ignore the agent-specific approach. It remains the most common vision of disasters, even in the work of social scientists. I think the other authors underestimate its powerful hold on disaster discourse, and how far it undermines the prospect of social understanding. As others have noted (Mitchell 1990; Wisner 1993), it is alive and well in the International Decade (IDNDR). That is focused upon "the impact of natural hazards" meaning earthquakes, floods, etc., and "sites where hazards are likely to strike most strongly." (National Research Council 1987: 10). Actions intended to reduce risk emphasize defenses against the direct impact of damaging agents, protecting or removing property from them (Castells 1991). The particular agenda for the UN Decade and related national initiatives, is to provide geophysical, technical, and organizational knowledge that "we"—the rich nations—supposedly have, to others who apparently lack them (Brook 1992). The decade was ushered in, academically, by some impressive agent-specific texts including overviews by Palm (1990), Smith (1992), Ebert (1988), and Alexander (1993); collections of process- or event-specific studies including El-Sabah and Murty (1988), Starosolszky and Melder (1989), McCall, Laming and Scott (1992), and Jones (1993); and a host of studies of specific geophysical agents as *hazards*.

Now, geophysical, technological and biological understanding or, as Dombrowsky (1995a: 252) puts it, "the auto dynamics of nature," provide essential insights for our work. The danger lies with, in Gilbert's terms, a "hazards paradigm": a viewpoint that classifies, explains and responds to disasters as if they were wholly or essentially a function of the agent that "impinges upon a vulnerable society" (1995: 236). This paradigm renders social understanding secondary, if not impossible, by placing the sources of risk literally outside society "in the environment," or presumed accidental unscheduled forces that "erupt" within. It has encouraged an adversarial view of the relations of environment and society—"the environment as hazard" (Burton, Kates and White 1978) or, as Gilbert (1995: 232) points out, as "enemy." Society—at least, communities, the public or populations—are made to appear passive victims of natural and technological agents.

The need to uncover the assumptions of the hazards paradigm arises because it remains central to the dominant view in this whole field (Hewitt 1983a). Other features of the dominant view make sense in relation to it. If people and communities at risk act at all, they are portrayed as doing so in ignorance, or out of mere perceptions of the hazard since technical expertise, not common sense or being there, decides knowledge. Solutions are found to lie in technical counter force. They need professionals and mission-oriented agencies to confront and tame nature. They must predict the extremes and target people at risk, informing and moving them around in relation to expert knowledge of the hazards.

The regime of mechanism and control

From a social perspective, a dubious but pervasive consequence of the hazards paradigm, is to encourage the reducing of danger and human responses to mechanism (Watts 1983b: 235). Disaster is conceived in terms of flow charts, or a causal chain, beginning with the impact of a physical agent, and moving along a series of pathways and response elements (cf. Hohenemser, Kasperson and Kates 1985, e.g., Fig. 11). The vocabulary is not merely one of geophysical and technological processes; of magnitudes, temporal or spatial frequency, polygons and zones. Persons, communities, and their concerns are also—and in this paradigm, perhaps, must be—reduced to mass, collective units, statistically distributed data points, and functions of abstract dimensions. Society is redescribed as (black?) boxes called population or GNP, public participation, and human perception processes or human wants, to take a few at random from texts on my shelves. These are linked by arrows in models that are, apparently, demonstration or organizational flow charts. They do not involve any phenomena and processes one would find in the real world. But they have taken on a life of their own. Somehow, it is the relations between such abstract entities whose stability, uncertainty or disorder is the issue, and in terms of other impersonal forces outside or running between them.

The result is an unsociable realm of impacts and energy, feedback and stability, information and control. Others have critiqued the scientism of this approach (Waddel 1983), but it is pervasive in positivist Western thought. I think that helps identify why the hazards paradigm has proved so robust a social construction for mainstream studies, and why we have to be conscious of its implications.

Most of the other authors use and take for granted a vocabulary of mechanism and control, and its underlying assumptions, if they struggle to improve its use. Dombrowsky's sophisticated arguments reveal and critique some important implications of the resulting technocratic constructions of disaster. The remarks on masking reality with false causality seem to parallel what I have just said. Yet, in the end he champions a search for the "algorithms that lead to disaster" (1995a: 253) and greater regard for the "autodynamics of nature (ibid.: 252)." Porfiriev builds his views unequivocally around mechanical and systems metaphors, and a central, defining notion of disaster as instability. It *is* a metaphor, and not only because of the compounding of material assumptions and value judgments. Mechanical systems are not unstable because they change, even destructively, but change because they had been unstable—which seems to reverse Porfiriev's usage.

Gilbert, who raises concerns close to my own, is nevertheless convinced that *uncertainty* is at the core of the problem (1995: 236). This notion is deeply rooted in the technocratic view of disaster as originating in unpredictable extremes, unscheduled events and uncontrolled agents. The hazards paradigm suits the fixation with geophysical or technological extremes, and a disproportionate technical effort directed at predicting statistically improbable events or thresholds. That also fits in with a managerial view of risk as probability (Whyte and Burton 1980). Uncertainty, at once defines an empirical and practical dilemma, excuses the failure to control, and justifies expert systems as alone able to probe and domesticate these further reaches of environmental and social "wildness" (cf. Hacking 1990; Shrader-Frechette 1991). One must question how far uncertainty is the appropriate factor for social understanding. It may be useful from the point of view of a government agency confronting mass, heterogeneous populations, to describe the incidence of rare, extreme events in nature or technological accidents. But it is a dubious description and placement of the persons, places, and life worlds of the rest of society! It is a strategy Dombrowsky (1995a: 244) describes as "cutting reality into parts which [may] fit into one's capabilities to handle them."

The most contentious result of the hazards paradigm generally, and the uncertainty focus in particular, is the tacit assumption of an unexamined normality; supposedly predictable, managed, stable and the basis of productive society. That goes along with the sense that disaster involves events having little or nothing to do with the rest of life and environment. Our job is then seen as to maintain, or to restore normal life as the polar opposite of disaster. Porfiriev, for example, sees the instability notion as important methodologically, in separating disaster from a routine or stable situation as well as other kinds of "extraordinary situation." However, if vulnerability is indeed important to the origin (causes) of disaster, then may not both routine and other forms of extraordinary situations be sources of his instabilities?

Extraordinary situations, if we follow Horlick-Jones (1995) line of reasoning, might well describe the times in which we live! Observe how some of the other problem "communities" find their concerns to arise from normal or everyday—if equally impersonal—forces such as overpopulation, debt, underdevelopment, rapid urbanization, pollution, global warming, militarization, and so forth.

THE CHALLENGE OF VULNERABILITY

Disaster and "normality"

The vulnerability perspective is strongly supported by Gilbert and gets an appreciative nod from the other authors. Vulnerability may be thought of as the other face of safety or security. However, it is less the literal definition of the term that needs stressing, than its place in the debate and the challenge to the hazards paradigm (O'Keefe, Westgate and Wisner 1976; Pelanda 1982b; Timmerman 1981; Hewitt 1983a).

True, vulnerability is being imported into natural and technological hazards studies, but to mean exposure to hazardous agents—"being in the wrong place at the wrong time" (cf. Liverman 1990). However, as Gilbert notes, more developed vulnerability studies "get rid of the overwhelming notion of 'agent'" (1995: 235). Rather, they place societal conditions at the center of the description and interpretation of danger. They also question a primary concern with disaster events, emphasizing the (permanent or changing) ecological context or Porfiriev's "routine" conditions. These are seen to decide the likelihood as well as magnitude of damages, whom and what they are more likely to affect, and peoples' ability to recover.

Vulnerability studies not only focus upon social conditions and ecological context, but tend to reject agent-specific categories and the separation of individual risks and practices. Real, public life situations are seen to require persons and communities to cope with, and in some sense balance, a range or mosaic of risks (Hewitt and Burton 1971; Quarantelli 1978; Ziegler, Johnson and Brunn 1983). In relation to that, any given response is situated within, if you will, a social economy of responses.

Social responses are often similar over a broad spectrum of dangers. That reflects the existence of diverse risks, many of which may be individually rare. Moreover, the social influence upon vulnerability to, say, earthquake, famine or toxic chemical risks, may arise in quite other areas of life. It may be largely due to housing problems, a megaproject, changing farming practices or the treatment of women. This is also another way of taking up the themes of "unnatural" natural disasters (Susman, O'Keefe and Wisner 1983; Wijkman and Timberlake 1984), the ecological meaning of technological dangers, and the sense in which all disasters or, at least, Quarantelli's "catastrophes," are complex emergencies.

Meanwhile, there is the question of whether responses to dangers can be reliable unless integrated into the fabric of social life. For societies, as opposed to science fiction, isolated or sleeping appendages, and remote agencies, that only fire up in emergencies, are not to be relied upon. That too challenges the focus on extreme events, accidents and expert systems.

Again, any societal decision to invest in particular risks or solutions affects other areas of society and risks. It may support them or deprive them of resources, create new dependencies or redistribute dangers. For example, Gilbert's "war paradigm" of emergency measures and attitudes to disaster is associated with and reflects the overriding commitment of modern states to military security. For them war is the greatest risk or source of disaster, military defense and research the solution.

The vulnerability paradigm provides a clearer sense of the idea of voluntary and involuntary risks than a hazards view. For these are identified with just and unjust social developments rather than impersonal forces and statistical profiles. A voluntary risk involves personal and group choices to take risks; or contractual and participatory arrangements where the client agrees to share some, at least, of the risk. The main sources of involuntary dangers are also in the social order and actions that place people at risk or deprive them of defenses and resources to cope. The emphasis is again on the social order, or human choice and action within it (White 1964), rather than mere mechanical processes.

Famine, perhaps more than any other form of disaster, has involved a powerful reappraisal of disaster through studying vulnerability (Dreze and Sen 1990; Bohle 1993). Of special note are the ideas of Amartya Sen, his co-workers and related discussions. They show how modern famines have not arisen from absolute food shortage within the food system or world to which the victims of starvation belong. Rather it is a matter of what Sen (1981) has described as *entitlements*, and of the capabilities and (dis)empowerment of those most likely to suffer increased starvation and uprooting (Watts and Bohle 1993). The Sahel famines became a major focus of struggle between those favoring a hazards (drought) paradigm, and others looking at social conditions. In due course, a transformed notion of the nature of famines emerged, recognizing the sociopolitical sources and complexities of who suffers and why (Copans 1975; Glantz 1976; Vaughan 1987). Reassessments of famines, famine relief and epidemics in India, in the colonial and post-colonial contexts, have also been a major focus of innovative and insightful social research (Bahtia 1991; Dubhashi 1992).

In the industrialized world, as Horlick-Jones (1995) emphasizes, and Turner (1978) traced so convincingly, the limitations and failures of the official and professional institutions for public safety are a focus of concern. Even more problematic, is the way they can be undermined by economic downturn and political change. The demoralization of institutions responsible for safety in declining or less prestigious industries, and their dangerous legacies, involve special risks (Ehrlich and Birks 1990). There was, perhaps, nothing at all the children of Aberfan, or their parents, could reasonably have done to avoid or prevent the coal tip disaster of 1966; or the poor families of Bhopal in the face of the toxic chemical cloud. But state institutions and inspectors, safety equipment and regulations, were supposed to take care of such things.

Such examples illustrate a double jeopardy of the modern world. There are so many risks over which people exposed have no control. But the safety experts responsible are not, and in many cases, perhaps, cannot be adequately equipped to do the job. Often, their effectiveness is an indirect consequence of the greater pressures to achieve production goals. Not only may that sometimes come into conflict with safety concerns, but gives little attention to the vulnerability of those at risk.

Challenging vulnerability

However, a generalized and abstract paradigm of vulnerability, is as unsatisfactory as the hazards paradigm. Vulnerability may be an unfortunate term. Unlike much of the work that it labels, the word emphasizes a "condition," and encourages a sense of societies or people as passive. Indeed, like the hazards paradigm, vulnerability can and increasingly does involve a focus upon the ways in which human individuals, the public and communities, are pathetic and weak. Latterly, vulnerability is being incorporated into the agendas of IDNDR in that way. It is being recast as yet another of the social pathologies like, or derived from, poverty, underdevelopment, overpopulation. Any problem or other setting that lacks "advanced" technological and governmental organization, and a strong consumer economy is seen to be prone to these "pathologies."

Gilbert (1995: 235) points out the ambiguities of the term and how vulnerability is also being identified in "relaxation of social and political bonds," with "uncontrolled processes" and "traditional criteria" in "complex societies." But the term itself hardly conveys the way people in nearly all circumstances, are active, creative and alert; or how organized decisions may endanger them by undermining their capabilities and resilience.

In fairness, much of the actual literature of vulnerability has emphasized human resilience (Timmerman 1981), adaptability (Davis 1981; Maskrey 1989); and the presence of risk aversion arrangements in all sorts of societies. This work finds persons and groups to be anything but mere play things of nature or fate. It leads to a critique of, especially, national and global strategies which generate vulnerability, or undermine people's capabilities to avoid or recover from disaster (Baker 1974; Morren 1983).

Wisner (1988: 16) argues that we will fail to get to the heart of the matter of risk unless we "create ways of analyzing the vulnerability implicit in everyday life." That is a basic challenge to both the hazards paradigm and abstract, systems thinking. Of course, Wisner (1988) and others with similar concerns, are well aware that everyday life is subject to unprecedented difficulties and changes, arising from economic globalization and deliberate intervention by the state. These have become critical aspects for everyday vulnerability. They can have a decisive influence upon whether, where and to whom disaster will occur. But this problem of the "everyday" raises some other social approaches and substantive issues that seemed quite absent from the other papers.

MISSING VOICES

Testimonies from the field

I was surprised by the way the other papers adopt a strictly Western, professional stance, concerned with general principles and abstractions, and with formal, essentially governmental organization. Perhaps it is the prevailing approach, but is it the only, or always the most insightful source of social understanding? Is it even the most relevant or best viewpoint for informing and improving administrative or organizational sociology? My own view is that it is not, least of all in the globalized context of IDNDR.

Meanwhile, some of the most important developments in the understanding of disaster in recent years come from workers on the ground. That includes those in the front-line of disaster response reflecting upon field conditions. Some are based on a rather specific experience with disaster relief and reconstruction (O'Keefe 1975; Waddell 1975, 1983; Davis 1981; Harrell-Bond 1986; Kent 1984; Hancock 1989). More broadly significant are studies by those who have spent extended periods in, and paid attention to the larger social and environmental context of the places where disasters have occurred. The vulnerability perspective has received its greatest impetus from such work in local communities. Those who speak the language and have some depth of knowledge of the cultural and ecological context, provide essential social insights. Examples have come from the rich countries, but usually disadvantaged groups and marginalized places within them (Erikson 1976; Shkilnyk 1985; Geipel 1982; Kroll-Smith and Couch 1990a). Most of the innovative studies and development of ideas are from work in poorer nations and so-called Third World or traditional contexts (Torry 1978; Lewis 1987; Liverman 1993; Winchester 1992).

The devastation at Yungay and in surrounding communities from the great 1970 earthquake in Peru, led to a series of extraordinary studies (Doughty 1971, 1986; Oliver-Smith 1992; Bode 1989). They arose, especially, out of a background of longer-term work in the region and extended residence in the area by researchers. They are longitudinal and comparative studies of these Andean communities in crisis, and their post-disaster reconstruction struggles. A primary concern was how the problems affected the survivors, appeared to, and were responded to by them. That provided a basis for a powerful reappraisal of disaster as a social issue (Oliver-Smith 1992).

Likewise, Watts' (1983a) work in Nigeria, served to ground a critique of existing hazards and disaster work, and a radical reformulation of the political economy of disaster. He combined a concerted investigation of indigenous famine mitigating arrangements, with analysis of what happened in the Sahel drought of the 1970s. Wisner (1993) situates the problem of disaster within his extensive analyses of underdevelopment and marginalization in East African rural communities, emphasizing vulnerability due to impoverishment of, and within, households.

Maskrey (1989) turns to a vulnerability perspective to show the advantages of a "community-based approach" to disaster mitigation. He provides examples of grass roots activities based on that approach. This, of course, goes back to the work of Davis (1981) on shelter provision and other relief and reconstruction activities. The emphasis is, again, on the concerns and culture of disaster survivors. When those are ignored by governmental or international agencies, this leads to a critique of the "disasters of relief". In that regard, I was surprised at so little reference in Kreps case study (1995a) to the tremendous ferment of ideas, the social activism, and critical literature surrounding the 1985 disaster in Mexico City (Robinson, Franco, Casterejon and Bernard 1986; Renée di Pardo, Novelo, Rodriguez, Calvo, Galvan and Macias 1987; Nunez de la Pena and Orozco 1988). If there was ever a case of a well-documented social catalysis precipitated by disaster and popular action, this is it. In his paper, Kreps (1995a: 259) seems to find the catalysis, and "elemental processes of the social order," to depend largely, if not wholly, upon how government and corporate organizations were mobilized.

Shadow risks and hidden damages

In recent years, studies of the societal underpinnings of disaster often identify silences in accepted or conventional work; voiceless and invisible presences; conditions and people ignored or marginalized. Issues are found to be hidden, masked, obfuscated or redescribed to suit other—also often hidden—agendas. More severely, there is Watts' (1983a) "silent violence" and the "quiet violence" of Hartmann and Boyce (1983).

The ideas of hidden damages and shadow risks have seemed quite appropriate to my own work in two seemingly very different contexts: geohazards in high mountain communities, and the impacts of air war upon civilians and cities. Vulnerability to, and harm from, landslides in a Himalayan village depends profoundly upon settlement and land-use geography, the social position and the life-worlds of people at risk. However, that tends to be hidden by the prevailing environmentalism of high mountain studies and most natural disaster work, as well as the assumed economic marginality of mountain peoples (Hewitt 1992). Again, the plight of civilians in wartime, or in an occupied city under bombardment, depends profoundly upon social geography and the position of civilians in the wartime state. That also tends to be hidden by the war-fighting preoccupations of air war and professional defense literatures, not least by much of civil defense literature (Hewitt 1987, 1994a; Macksoud 1992).

This sense of revealing what has been hidden is perhaps inevitable for studies that confront dominant paradigms which ignore their concerns. However, the hidden damages and shadow risks do not refer to superficial or secondary items. In that sense, the analogy with the ideas of hidden economy and shadow work is appropriate. The latter identify large and pervasive features of economy and labor, of gender and status, which are hidden or placed in the shadow by mainstream ideas of political economy. That leads to a critique of the discourse, as much as an identification of concerns.

The way persons and issues are hidden and reduced to other terms in modern contexts, may be another way of coming to the existential problems of risk, trust, and psychic survival that Horlick-Jones (1995) raises. I am not sure, however, whether he also sees these as impersonal, inevitable consequences of modernity—fundamental aspects of the fabric of modern societies. Perhaps they are, but is the fabric a product of itself, or the ideology and actions of the more influential persons and institutions? Shall we say the bombing of Hiroshima, the explosion at Chernobyl, or the presence of those particular buildings that collapsed in Mexico City in 1985, ultimately could not be avoided? Arose out of inherent flaws and disorder in the modern human condition? Such a construction is laughed at by scientists as "fatalism."

Was no one, and no pattern of human decision and regulation, responsible for the Holocaust? That archetypal juggernaut and outrage of modern times, is the subject of some profound reflections upon disaster and modernity. They include a literature that insists upon not losing sight of the faces of the victims, nor the perpetrators, when examining "the system" that destroyed so many lives. Among other things, the destruction of European Jews depended upon a massively orchestrated and cynically "hidden," but deliberate, manipulation of the science, bureaucracy, and infrastructure of a modern state (Furet 1989). Auschwitz has been called "a monument of science and technology" (Muller-Hill, quoted in Lifton and Markusen 1990: 103); the extreme projection of discipline, routine, and the regime of production schedules (Kogon 1958). Others refer to the banality and "normality," as well as the uniqueness, of the Holocaust (Bauman 1989: 83).

These and other disaster events and places of atrocity may not arise in deterministic ways or out of the decision of "one man," nor, except rarely, from conspiracies. They do seem to be ushered in by the ways in which decisions are made, practices guided, and concepts *socially constructed*. In the modern state, these turn calamitous for many, in the absence of the accountability of power to an independent judiciary, press, art, and academia. They, in turn, need to be responsive to and part of civil liberties and the subordination of institutions and ideas to the concerns of persons and communities. Decisions made to suit the agendas of powerful institutions, or ideas that construct people as instruments and abstractions, refusing to see their faces or feel their pain invite, if they do not depend upon, widespread cynicism of dominant groups and agencies (Castel 1991; Sloterdijk 1987). What, then, shall we do with a disaster sociology that sees solutions only through technocratic systems and strategies; or disaffection in terms of their failure to be trustworthy?

This problem of not losing sight of the faces and communities of risk leads to something else that is absent from, if not demeaned by, the dominant literatures. It is personal testimony, and from the victims of disaster. That involves the methodological challenge of work that pays attention not just to the local conditions and communities, but to the voices of persons living there.

Whose testimony?

In many ways, the work of recovering and listening to oral testimony from the victims of calamity—"the voices of anonymous suffering" as Meyer puts it, (Meyer and Poniatowska 1988)—goes to the heart of the problem of finding the human and the social in risk and disaster. It, alone, provides a means to obtain adequate witness to the conditions of danger, just who and what has been hurt, and their needs. How else can the academic ear be opened to the worlds of ordinary people, children, the elderly and infirm, those from disadvantaged groups and other cultures? One group of oral historians reflecting on the absence of women's views asserts: "when women speak for themselves, they reveal hidden realities . . . new perspectives emerge that challenge the truths of official accounts and . . . existing theories" (Anderson, Armitage and Wittner 1987). Even if just partly valid, this has major implications for social understanding. Meanwhile, some of the most innovative and widely praised studies have been based largely on the testimony of survivors. Lifton's (1974) study of Hiroshima and Erikson's (1976) of the Buffalo Creek flood come to mind. Again, oral testimony identifies the moral and humanitarian as well as social perspective on the space of harm: "oral history not only shares the fate of the vanquished, but is also born at the moment of disaster and of collective social forgetfulness" (Meyer and Poniatowska 1988: 15).

Letting those in hazard speak for and of themselves, is one of the few possibilities for keeping the faces and pain in the foreground of interpretation and response; as part of the social evaluation of problems and responses, not merely as advertisements on the front cover. The battle against "disaster pornography," the exploitation of tragic images and heartrending words by officials and media for the promotion of their own organization, must also be fought in this area. However, from a scholarly perspective, a person's words are "evidence" already verbalized and communicative. The testimony cannot be merely "subjective," or it would be gibberish to all others. Its intersubjective nuances and symbolic meanings do, of course, require careful and cautious attention. To listen to, value, and try to understand the plight and experience of ordinary people in everyday settings, and the victims of disaster, presupposes a concern with who they are and where their experiences take place. To focus on their words is to recognize that these are the only way to recover experience in other places and times. To pay close attention to what they say, their story and concerns, gives them direct entry into the concepts and discussions of social and disaster research. This is an essential step towards giving those who do not get published some participation in and control over the impersonal processes and citadels of expertise that tend to dominate the disaster community.

These concerns become all the more important when we see the bulk of disaster work actually moving in the opposite direction. It is increasingly global. Its agendas are set by state and international organizations. We use ever more machine-mediated technical systems and human concerns translated into the abstract languages they work in. Every center, university, and agency displays an obsessive following of fast-breaking scientific discoveries, information systems, and networking; the fashionable words and debates in the wealthy nations. Professionals may, perhaps, not get anywhere without having some involvement in all that. And I am no less beguiled by these "new toys" than my colleagues! Yet, the more technological and abstract our work environment becomes, surely it is the more important to get out into the field and confront our abstractions with evidence on the ground. How else can study recognize, let alone comprehend and work to assist the realities of people's lives? (Hewitt 1994b).

Meanwhile, even the most downtrodden and impoverished persons are, and to live at all must be, active, intercommunicating, socialized and socially responsive beings. The least of them have huge untapped abilities, as well as a capacity to survive where many of those who govern or study them would not. That applies to individuals, families and, outside the most atomized of Western social settings, in communities and cultures. If people stop being active, or are prevented from developing and acting according to the capabilities and values appropriate to their contexts, then things indeed fall apart. This leads to the last but underlying theme of this commentary.

GOVERNMENTALITY AND CONTROL

Missing agendas

Unconsciously or not, the other papers seem preoccupied with how disaster and society look from above, abstractly, and at a distance. Except for the latter part of Gilbert's and Horlick-Jones' papers, they are not concerned with conceptual issues, or the results of empirical and comparative perspectives, such as are found, say, in the classic work of Walford (1879), Prince (1920), Sorokin (1942) or Barkum (1974). Rather, the question behind the question seems to be: How do we characterize disaster as a social problem for centralized organizations and professional management?

It is as though we have been asked to imagine we sit at the control panel or on the board where impersonal systems are administered, and conditions require a strategic vision and interventionist mission. This is neither an invalid nor an unimportant point of concern. Given the context in which most of us work, are consulted, obtain research funds, find employment for students and trainees—Gilbert's *market*—it may be *the* question. It is the overt and foremost preoccupation for Porfiriev's paper and, to a large extent, of Kreps, so perhaps they are justified in not giving space to other issues. Gilbert's "third paradigm," does seem to question the relationships of government, communities, and social action (1995: 236) , and Horlick-Jones (1995) considers their apparent unraveling. Yet, this directing of all the discussion to technocratic and organizational problems, results in a rather limited sense of the social content and the problematic of disaster. There is also another taken-for-granted social construction: the way state formation and governmentality have shaped notions of public safety and the practices of civil security.

The disaster community and field has taken shape largely as a response to the recurring failures of centralized safety measures and social control. Two major issues or constraints have shaped the preoccupations of disaster studies. The first, as emphasized by Horlick-Jones (1995: 311), involves questions of *control*—or rather, the occasions when it breaks down or is seen to have been lacking. The second, is a construction placed on events in terms of the practices and powers of centralized organization.

In such terms, a basic but unspoken context for the disaster community is the other domain(s) and levels of risk in modern societies, sometimes called chronic or lifestyle hazards. They seem different from our more extreme concerns, usually being of dispersed, if widespread, incidence. What links and distinguishes, perhaps contains these socially, is *routine* treatment. And what is most relevant to the shape and role of our field in modern societies is how such dangers are the subject of expert treatment in asylums, courts, prisons, clinics, war colleges, etc. They are identified with prominent fields of study such as medicine, criminology, actuarial science, engineering safety, and military history. To the extent that these permanent or chronic dangers are well understood by established fields, and well-managed by responsible institutions, our field does not consider them, or at least, not as matters requiring special attention. That is so, even though the chronic dangers and damages involve by far the largest privations, material losses, and untimely death, and by far the largest investments in public safety and national security. Normal life, in such a context, is not so much ordinary, everyday, or safe, but obedient to large-scale goals, regulated and disciplined centrally.

The fields and institutions that ensure such well-orchestrated, predictable existence in the mass have, by default perhaps, come to frame the special concerns and responsibilities of disaster studies and agencies. We are not just pressured to address problems that are extremes or emergencies. Disaster threatens and involves destruction or disintegration of the extensive, orderly patterns that bind together the large space and many places of modern material life. Disasters are problems that are, by implication and in fact, out of control, in that they break out of the modern mold, or challenge its effectiveness. That is how the tell-tale *un*-words seem so readily to define our concerns—the language of the *un*anticipated or *un*scheduled events; *un*certainty and the results of accident, human error, bad design or underdevelopment (Hewitt 1983a).

In this context, the social problem of disaster is not primarily one of, say, crisis, devastation, extreme experience, and emiseration, let alone of tragedy, violence or misrule. It is all about (loss of) control—meaning control, as Horlick-Jones (1995: 311) suggests, within a particular kind of public order. Problems are constructed as the absence, limitations or failures of 'police' and 'discipline', as defined by Michel Foucault, and as demonstrating the need for more of them (Foucault 1975; Arney 1991; Burchell, Gordon and Miller 1991). Police, here, means not just crime prevention, but all forms of regulation and enforcement of state powers. Discipline refers to all the ways in which individuals are trained and required to behave in impersonal organizations.

These notions describe, better than most, how science and administrative technologies have been deployed by modern states in the fields of disaster reduction and emergency measures. I suggest that these questions of governmentality identified by Foucault and others, have a major bearing upon the views of disaster expressed in the other papers, and as the basis for a critique of a prevalent form of organizational sociology directed at improving modern state administration and its approach to social problems. The fascination with the language of mechanism turns out to be not so much scientism or aping the hard sciences, but a creative encoding of the ends and workings of this kind of modern social organization. Dombrowsky's paper is particularly relevant here, and his comments on "triggers" that call forth appropriate tool boxes.

It is not surprising, as Gilbert notes, that civil defense provides the model of organized disaster preparedness. It extends the demands of national security to civilians. The first official concern of the authorities and a large role of the military and police services in most disasters is to restore order. And that means the instruments, infrastructure, and appearance of a *centrally administered* order.

"Disaster sociology" for whom?

Most people working in this field have taken it for granted that, in some sense, modern science, our administrative and legislative systems, offer the only or most likely way to disaster reduction. Very few question the dominant view, or that solutions must and will be found in technical approaches. The arguments of those who do doubt that appear increasingly relevant in non-Western or traditional settings, and in relation to the plight of disadvantaged persons and groups generally. Yet, no one can ignore modern institutions and systems. Only if these are successfully confronted and influenced will alternative approaches have a chance to flourish.

But where does organizational sociology lead when it takes improving the effectiveness and centralized administration of agencies and expert systems as its focus? Can it offer any advice, other than a more totalizing penetration of government and powerful interests into everyday life, greater surveillance and militarization of public and private space?

The modern history and lessons of public safety and security seem to me to be at least as much about the dangers of excessive state or authoritarian control as too little. Risks from dangerous enterprises, for most people, turn out to arise as much from the efficiency and power of technocratic, impersonal institutions as from their failures. The gravest developments in insecurity, untimely death, uprooting, and impoverishment of peoples in our century have derived from totalitarian violence or monopolistic deployments of science and technology and a war paradigm. They have continued to dominate the landscape of disaster during the first half of IDNDR. There is no indication that such grave dangers arise from too much civil freedom, high quality and equitably distributed social services, devolution of powers, or traditional arrangements for social security and disaster response.

None of this is to imply a rejection of social security measures or aspects of public welfare and emergency measures as the proper duties of government in the modern state. It is a question of whose concerns, what sort of societal evidence and contexts should inform such arrangements. Who, in a civil society, should be serving and policing whom? Whose interests should inform and decide the role of international humanitarian and disaster mitigation efforts? This could be a way of seeing the terrain of Horlick-Jones' concern with public trust and sense of betrayal (1995). However, I did not sense that the other papers were worried about these things, or saw them is an important issue for social understanding and disaster reduction. Perhaps I am mistaken, but they left me with an overwhelming sense that organizational sociology is the central issue, and its proper sphere that of technically and centrally controlled administrative measures.

REACTIONS TO REACTION

In the following section, the four authors of the chapters to which Hewitt reacted provide their own reactions.

7

REPLY TO HEWITT

Claude Gilbert

Hewitt's paper aims at summing up all the articles written in the IJMED volume issue to answer Quarantelli's question: What is a disaster? At the same time, it introduces a critical stance regarding them and insists on the inclusion of a certain number of perspectives *excluded* by these works. As the other articles are not in my possession, this reply is only an answer to Hewitt's lecture on my own paper. Its main goal is to try again to contribute to the definition of what a disaster is, and how it can be viewed in the field of social and human sciences. This paper will not develop a point-to-point answer to Hewitt's article, but will rather try to state more precisely my personal point of view following his reasoning.

The first point underlined by Hewitt concerns the conditions in which knowledge is elaborated about what we are used to calling "disasters." In my earlier paper, I mentioned three points: the conditions in which the question of disasters comes to be "written on the agenda" and categorized as relevant to the social and human sciences, the weight of research sponsors in the definition of the object of the study, and the boundaries that limit the comprehension and the perception of this objective given the "cognitive resources" of researchers, and so on.

Raising such questions was certainly not a way of doubting or criticizing the notions, the concepts, or the knowledge that have been developed until now. These kinds of questions can usefully help analyze the work of researchers from a sociological point of view, following what the sociologists of science do, in many other fields. Like other notions used in sociology, the notion of disaster has emerged in a particular conjuncture that has to be taken into account when studying the genesis of "disaster" as a conceptual construction, with all the constraints that can be associated with that particular moment, especially the intellectual constraints. Disasters as an object of study are clearly and strongly associated with the Cold War, with the traditional assimilation to wars made by administrations (simply, for example, to try to understand the reactions of people to bombings) with studies on disasters demanded by military administrations in the social and human fields. This context has strongly conditioned the definition of a disaster; a clear indication of this is given by the high number of military metaphors used in those studies. The problem that researchers had to solve on the institutional, financial, as well as theoretical, levels, and that they did solve in different ways, was to "take position" with regard to this founding heritage. However, this is a usual problem found in the human and social sciences in general, and it has therefore to be treated not from a "moral" or "ethical" standpoint, but from a strictly sociological one. To illustrate the point with Hewitt's favorite example of "victims," what is the meaning of this notion—that has emerged recently in France—in the field of risks? To which conjuncture, which kind of constraints, which type of *market*, as Hewitt puts it, insisting on the term I used, does the production of this notion refer? On each one of these aspects, such a question must be kept in mind.

The second point introduced by Hewitt's remarks concerns the nature of a paper similar to the one I presented in an attempt to identify the main paradigms or, to put it more simply, the main "sets of propositions" that researchers utilize when discussing disasters. Even though my preferences and my choices appear in my paper, I positively wished to put a critical distance as regard to each paradigm and each set of propositions in order to outline what each one of them develops, underlines, or conceals. For example, in accepting an agent-dominated perspective for the analysis of disaster, one is then obliged *ipso facto* to reason in terms of war, in terms of aggression and defense. This is a statement—we can take as a fact of rhetoric—and in principle, in my effort to clarify concepts, neither did I support nor criticize this approach. I simply observed that this kind of reasoning is frequent and easy to use in that it allows researchers to talk the same language as the sponsors of the studies, and it can easily be translated and reformulated for "communication needs." I also observed, on the other hand, that this way of reasoning establishes an asymmetrical situation between the agent, recognized as the active principle, and the population, presented as the passive object. In this kind of configuration, the core of the disaster is systematically situated outside the collectivity and this directly results in dodging any questioning about internal vulnerabilities. I just noted this situation and did not want to say if this paradigm is relevant or not, adding however that its use seems to have diminished in recent years compared with previous decades.

Likewise, when referring to certain notions that allow the construction of paradigms, I only gave an account of their uses and the ambiguities related to them. For example, as Hewitt says, I noted that the notion of *vulnerability* includes many different meanings and is used rather imprecisely. But this ambiguity explains the over-utilization and the richness of the term. When in France a book is published entitled *La Société Vulnérable* (Theys and Fabiani 1987), a consensus emerges among several actors of the political world, the administration, the mass media, and university and research sectors about vulnerability. All these actors share a common opinion on the central place it occupies in the problem of risk-solving, but they do not have the same definition of the notion itself. Rather than criticizing this ambiguity, I just observed that, when we introduce this notion into the debate, we automatically put the question of disaster at the very heart of human collectivities. In my perspective, this shifting is in itself meaningful.

Likewise, in talking about the question of *uncertainty*, I just mentioned the frequent use of this notion during the last ten years with reference to disaster. And there again, I neither expressed agreement nor disagreement. The introduction of this notion just let me underline that the question of cognitive problems, the loss of reference points, is becoming more and more important in the analysis of disasters. This does not mean that in my personal works, I make the assimilation between normal situations and certainty, as Hewitt seems to believe. If I had to formulate a judgment, I would say that normal situations coincide, through setting up everyday routine, with a permanent fight against uncertainty, and also against "complexity" (see Rosset 1979; Latour 1995). This debate, although necessary, is however different from the point I made in my text that was intended to try to identify which "rhetorical resources" researchers can use to analyze disasters, and also the kind of constraints they have to face within each of the paradigms mentioned.

Finally, the third point made by Hewitt concerns what he calls the "excluded" or "forgotten" perspectives. Trying to sum up his reasoning on the basis of the other papers he kindly sent to me, I would say that his purposes may be synthetically enunciated as follows: the ordinary citizen's opinion should be reintroduced into the analysis of disasters, instead of having it confiscated by the authorities—the administrative, political, as well as scientific, authorities. This pattern is essentially similar to the "fight" he has been carrying out in order to have the victims of World War II bombings taken into account in the analysis of the war itself. Hewitt's position is at the origin of several remarks on my part, which may appear contradictory.

First, I totally agree with the idea that disasters have been considered as the affairs of the public authorities rather than the affairs of citizens and, in doing this, the perception of "the" population as a whole is merely passive and bound to be directed and commanded in cases of disasters. This kind of talk is common and could have offered new subjects of discussion for Michel Foucault (1975), also mentioned by Hewitt. In any case, it is rhetoric and, in France particularly, it is the rhetoric of legitimization as a great part of state legitimacy that is formally based on its capacity—recognized and attributed—to "face" national calamities. Actually, when one observes what really happens in cases of great disasters, it is clear that there is a gap between all the possible hypotheses—paying no attention to elected representatives and citizens—and what happens on the ground. All the recent studies carried out on deep crisis situations in France show, like all the situations studied by the Disaster Research Center, that human collectivities do participate in the management of disasters. It is well known that in case of earthquakes, such as the one that happened in Mexico City, the assistance to the victims comes first of all from the other survivors, with the means used by the authorities contributing to the emergency only to a very little extent. In short, what is interesting in the empirical studies of disasters is the surprising capacity of the reaction and self-organization of people outside any usual public or institutional structure. In other words, the notion of "victim" here is very far from what is usually conveyed by the mass media, nongovernmental organizations (NGOS), or by major international organizations.

By admitting this—and I think that here is a question that Hewitt indirectly asks—it is necessary to ask why, even when the capacity of citizens to react formidably in case of disasters has been acknowledged (Quarantelli 1986: 87), there are so few works, analyses, and research on this aspect compared with the very high number of investigations on the functioning, good and bad, of public powers and official emergency systems. It is deliberately concealed. But to understand and analyze this phenomenon, however, is it necessary to denounce a sort of "conspiracy" by political, administrative, and scientific authorities monopolizing the question of disaster? I am not sure of that. In a sociological perspective, we can, on the contrary, try to understand why in the field of disaster management, public authorities in several countries frequently assert that they are able, against all evidence, to have everything under their control and order. We have also to ask why the behavior and reaction of citizens, to which Hewitt pays particular attention, are minimized and left out of consideration. I think this kind of question really is an interesting point for research. But instead of having a "well seen" or "correct" rhetoric on victims and the power of public authorities, we should try to analyze: (1) what is it that goes again the integration of citizens to the set of actors concerned, and (2) what does it mean today that a sort of revindication appears to acknowledge and to integrate citizens within the realm of actors in disasters.

In conclusion, I could say that Hewitt's remarks are very useful, even when not exactly formed. They oblige one to revisit and rethink incomplete propositions. In my personal case, they have convinced me that it is necessary to insist overtly on the following point: the study of the state of knowledge as tightly linked to the conditions in which it has been produced, does not necessarily imply the refutation of that knowledge. Any knowledge is produced in a conjuncture that heavily determines its limits and its possibilities. Hewitt's remarks were also the occasion for me to make precise the nature of my paper and of my effort to try to make a distinction between the main paradigms utilized in this field, and efforts of putting them to the right distance. Finally, Hewitt's accent on ordinary citizens, the victims, obliges a reflection on the deficiencies existing in the field of disaster analysis. It is up to him now to revisit his own analysis on the basis of our "bad remarks and questions," rethinking over some initial presuppositions of his analysis. This could be a useful way of progressing.

8

DEBATE—TEST—DUMMY

Reaction to Hewitt

Wolf R. Dombrowsky

There are many famous last words. My favorite one was said by Debate 4, a word-processing machine constructed for the simulation of scientific arguments. "Missed the mark" was the last phrase it uttered before being disassembled. Later, an autopsy showed that the argument recognition system was disconnected. Thus, the programmed aim, a scientific debate, could not be ignited because the machine reprocessed its own database.

That was exactly how I felt when I finished reading the Hewitt paper. His nice and friendly cover letter accompanying it, initially induced the opposite feeling. With his sentence in mind—"I hope you will find the comments that do relate directly to your paper are fair and constructive, and that the paper will at least provoke some useful debate," I looked forward to reading his paper. I longed for such a debate. From my point of view, our field of study badly needs a debate. There should be one between the various scholars and their different paradigms, and between the views held in different countries and their adaptations for a disaster sociology. From the first time I heard it, I liked Quarantelli's idea of having a cross-cultural collection of theory-oriented papers on the question: "What is a disaster?" I was very enthusiastic about his idea of inviting a discussant to write a reaction paper with the possibility also of later replying to his remarks. This was a good chance and a good challenge. Theorems that might seem to be large in Germany might shrink to being very small when they were put into the intellectual context of the North American tradition of disaster research. There would be a chance to learn and grow. Conversely, some dwarf ideas might mutate into Davids who would conquer Goliath. A great challenge for paradigmatic change and progress. That is what I looked for. What did I find?

After reading Hewitt's paper, I was disappointed and disillusioned. My illusion had been to expect a debate. Instead, Hewitt wrote an interesting and agreeable article. However, his article has only very little to do with the course of the arguments in the five initial papers from the roundtable. The papers are mentioned, of course, but more in the sense of being used as illustrations rather than being critiqued. There is no attempt to rethink the train of thought in the five articles, to test their load capacity and suitability in terms of building theory, disaster theory.

The basic structure of our "dialogue" had been arranged in this way. I talk about teeth and Hewitt asks: "Hey, what about the toes?" I try to develop logical and epistemological arguments and Hewitt says that I have neglected famine and the food system, the International Decade for Natural Disaster Reduction (IDNDR), and the insights from relief workers in the field. As a matter of fact, my institute is a member of the German IDNDR committee. I have done research and on-the-scene counseling in Rwanda/Zaire and other dangerous places, and we continually train and educate field workers. However, all this has nothing to do with my paper. It does not deal with the IDNDR, refugees, plagues and famines, relief workers and missions, and it was never my intention to do so.

In addition, besides a general shadow-boxing, there is what might be considered a punch below the belt. Hewitt tries to impute a hidden line of *Weltanschauung* in the five papers. According to him, disasters are too much seen as a problem of management organization, power and order. Very easily his argument shifts from technical to technocratic, from instruments to instrumentalism, and from managerial to being cold-hearted and inhuman. Consequently, to him, Chernobyl, Hiroshima, and Auschwitz appear as the logical outcome of a thinking that is identified as being in the five papers as well. That is far beyond shadow-boxing.

To sum up, if I were to discuss Hewitt's contribution in the same manner he used with our presentations, I would need only to merge the five papers under a common title: "Excluded perspectives in the social construction of disaster." In doing so, each party would then orbit in their own circle of autoreferentiality (as indicated in the system theory of Niklas Luhmann), that is, in one's own sophistic thinking. The idea: if I were to do as Hewitt did, we would all circle in our own thinking, and instead of dialogue, insults—such as references to Auschwitz, Hiroshima, and Chernobyl—would be at least implicitly hurled. Given that, a debate would never start, but the dummies would die while muttering "missed the mark."

9

A REPLY TO HEWITT'S CRITIQUE

Gary A. Kreps

In offering the following brief response, I will comment only on Hewitt's critique of disaster research as it relates to the substance of my paper. None of the authors, except for Porfiriev, reviewed the other papers in writing their own. Thus, while each of us would probably take issue with at least some of what the others have said, we have only been asked to respond specifically to Hewitt's comments. Actually, Hewitt gives only passing reference to any of the papers in his more general commentary on disaster research. By very clear implication, however, my work is characterized in a fashion (as an exemplar of organizational sociology) that I think is mistaken and in need of correction. I will therefore summarize Hewitt's general view of mainstream disaster research, the mistaken implication he derives about my own work, and the corrective that is needed. I do think, however, that Hewitt and I share the same general goal for disaster research, namely, more powerful descriptions and explanations of disasters as sociological phenomena. People at risk do matter, and we need to study them on their own terms.

If I understand him correctly, Hewitt believes that the social construction of disaster is dominated by an agent-specific, mechanistic, impersonal, and essentially amoral research paradigm. The epistemological foundations of that paradigm, he argues, can be reduced to an unquestioned faith in science and rationalistic problem-solving. As applied to public policy, he suggests, the central analytical tools of that paradigm are formal organizational models of disaster prevention, mitigation, and response. The major purveyors of that paradigm, he contends, are disaster researchers from various disciplines, government agencies, the power elite, and even the general public of industrialized societies.

Hewitt then argues that what is overlooked by those wedded to the above dominant paradigm are the unique histories and cultures of real people throughout the world who are in harm's way. What he believes should be done is to give these threatened people an active voice in things such as the

International Decade for Disaster Reduction. How? The answer for Hewitt seems clear enough: the dominant paradigm that is so characteristic of Western societies must be transcended, if not rejected outright. It should be replaced by an interpretive (as opposed to positivist) research paradigm that takes seriously the perspectives and actions of threatened populations more directly into account—a sort of grass-roots approach to building a more analytically rich and normative organizational sociology. Such an alternative paradigm would compel an interpretation of disasters as products of ongoing practices, social discourse, and ideology. The *raison d'être* of this alternative paradigm would be to identify social actions that are just, i.e., public policies that speak to the needs of real people who are threatened or actually victimized by disasters.

While my paper is mentioned only a couple of times by Hewitt, it is clear that he sees my work as an exemplar of the kind of organizational sociology (and therefore the dominant paradigm) that he wants to replace. Now there may be a sterile organizational sociology out there, but I do not think it dominates sociological studies of disaster. I invite Hewitt to look carefully at my own research program with Bosworth as but one example among many. I think he would find us to be very sensitive in our work about his admonition to "get out into the field and confront our abstractions with evidence on the ground." Indeed, that is precisely what we have been doing with our field studies and archival analyses for many years. The results, we believe, are more powerful accountings of disasters as sociological phenomena (Kreps 1985; Bosworth and Kreps 1986; Kreps 1989b; Kreps and Bosworth 1993, 1994).

Having stated that, we affirm once again our long-standing commitment to abstract theorizing and basic sociological constructs. We have found such constructs as formal organizing, collective behavior, and role enactment to be excellent analytical tools for capturing the social construction of hazards and disasters. We also affirm our commitment to fundamental social science. It forces an accounting of the social construction of disaster that is at once rigorous and illuminating. We agree with Hewitt that people at risk are both objects and active subjects of sociological inquiry. The dialectic of action and order is how the social world really works. In trying to unravel that dialectic, we therefore are equally attentive to collective behavior and the activities of formal organizations. Indeed, the circumstances we study force us to recognize that formal organizing and collective behavior are two sides of the same coin. Similarly, our data show how and why structural conceptions of pre- and post-disaster roles must be informed by symbolic interactionist principles. Of course, the retrospective case studies of an earthquake and a power plant accident can only illustrate the life histories of disasters. I agree with Hewitt that the major insights come from the details of what people on the ground are doing.

I do differ vigorously with Hewitt in one sense. I am not preoccupied with making normative judgments about the strengths and weaknesses of

Western or any other societies. There is a place for such judgments in social theory, but that is not what my research program has been about. Hewitt would undoubtedly fault me for that omission, but that does not make me an apologist for the so-called dominant research paradigm that he finds unacceptable. In the end, Hewitt has created a caricature of disaster research in order to make a point at the expense of many colleagues—colleagues who are no less committed than he is to developing powerful descriptions and explanations of disasters as sociological phenomena. I respectfully disagree with both Hewitt's strategy and the major thrust of his critique.

10

WHAT I SEE IN HEWITT'S MIRROR IN REREADING MY PAPER

Boris N. Porfiriev

I really do believe that for someone to understand the pluses and minuses of an original conceptualization or interpretation of a phenomenon or a process (e.g., a disaster), the best and only way is to bypass the review of a peer, whatever that may be. Fortunately, even a damning criticism is not a final diagnosis of the inappropriateness of one's theory, and luckily I as an author do have a chance to respond. Moreover, the latter possibility means that the critic might find himself in a vulnerable position, and at least have a chance to personally try the fitness of a vulnerability concept of disaster.

This last is a half-in-jest statement, but I do want to indicate that I have a genuine respect for Hewitt's valiant attempt to scrutinize and comprehensively reflect upon five different papers of five different authors from five different countries. I take his comments as real reflections. They look like speculations and reasoning from what he found in the papers, and present amendments to what, in his view and rigorous critique, was insufficiently discussed or even missed. And that is why I consider his comments as both useful and stimulating. However, this does not mean that I completely agree with everything written by Hewitt, so I will try to comment briefly on his critical review.

I will confine myself to two things: first, I will touch only on my paper, and will not deal with the other papers; second, only major incoherent or unclear positions will be discussed.

The most important thing about the last point is that it refers to the intrinsic characteristic of disaster as a conceptual category. Hewitt considers that I built my views "unequivocally around mechanical and systems metaphors, and a central, defining notion of disaster as instability." He then says that this is a metaphor partly because: "mechanical systems are not stable because they change, even destructively, but change because they had been unstable." That reverses my logic. While I completely agree with my critic in noting that my concept of disaster is built upon the *system* approach (but not "metaphor"), I cannot but disagree with his point about "unequivocally mechanical metaphor" and his speculations that follow on the changes and stability of mechanical systems.

First, in mechanics and mechanical/electromechanical systems, e.g., a watch or personal computer, which can serve as good representatives of a fine and complex system, one of the main principles or prerequisites of their functioning is a close contact and interaction between the components or elements of such mechanisms. This gives grounds for specifying that these are deterministic systems, which "unequivocally" differ from more flexible social systems that exist within stochastic domains, and that is the focus of my writing. From this viewpoint, my approach looks as if it corresponds well with Claude Gilbert's conceptual model that is based on an "uncertainty" category, which was however also the subject of Hewitt's critical arrows.

I believe that my respected critic was misled by the basic category I introduced while reasoning within a system approach about disaster, i.e., "stability." Hewitt labels it as a mechanical term and speculates on changes and stability relationships within a mechanical system, and correctly points out the possible destructivity of such changes that could mean eliminating the system as a functioning entity. For example, breaking even the smallest (elementary) part of a watch or a personal computer means losing the whole piece, though its shape may be preserved. That is absolutely clear. A mechanical system involves a determined causal process that intrinsically lacks the factors of voluntariness and acceptability of risk—factors that serve as one of the basic watersheds between mechanical and social systems in general.

What I tried to argue is quite different and referred to stochastic phenomena and processes typical of social systems. I am talking about the "strategic stability" that is kept by a social or sociopolitical system (e.g., community, society, civilization) even in case one of the elements or subsystems is lost. More explicitly, the "breaking" of a person's life or the life of a family which constitute the basic elements of any society, though it is real trouble for them and their closest relatives, however, does not mean the disappearance of a community or society. It will prepare and strive against the impact of "specific agents," if risk is involuntary and/or mitigated or even eliminated in case the risk for the community is voluntary in nature, thus approaching socially acceptable limits.

In case it succeeds, the holisticity of a social system, being the cornerstone of a system as such, is not broken (although impinged) and preserved due to the keeping of a principled balance between and inside sociodemographic groups; links between other individuals and families; functioning productive facilities and transport, etc., which represent what I call "strategic stability" or more briefly "stability." It may be lost in disaster when there is a "breaking of collective routines of social units" (see my article), not of individual units, thus aggravating the vulnerability of the social system.

With this latter notion, comes perhaps another unclear position of mine. Hewitt expresses strong doubts about my separation of disaster from a routine or stable situation, as well as other kinds of extraordinary situations. He first asks "if vulnerability" is important to the "origins" (causes) of disaster, and then asks if both "routine" and "other forms of extraordinary situations" could not be the sources of "instabilities?" To be sure, the mentioned factors or circumstances may be and really are the sources of or catalysts for deepening instability, thus accelerating or aggravating the situation in a major disaster. But paraphrasing the well-known question: "How safe is safe enough?" to "How much stability is stable enough?", I would like to point to the key issue of the relationship between necessary and sufficient prerequisites for a situation to be characterized as disastrous. I consider that the factors mentioned by Hewitt, while citing me, are among the prime ones, but surely they do not constitute a set of circumstances self-sufficient for a disaster. In short, it is a problem of defining thresholds or weighing risks while engaged in decision-making. But I would not like to go further considering that the basic question asked by the editor of this special issue, Quarantelli, which I honestly tried to answer was: "What is a disaster?" and not "Why a disaster?"

The last comment I would like to conclude with concerns the human dimensions of disaster conceptualization. Undoubtedly, the aforementioned theoretical constructions, as well as my original paper, lack "living" pains and fears, and the sufferings and tears of real people trying to cope with the realities of a disaster. In this light, all the speculations and reasonings extensively discussed by Hewitt in the bulk of his critical review, should be considered as fair. At least I do, since I respect the famous saying of Dostoevsky that a child's teardrop outweighs all material gains.

However, whatever the models or conceptualizations are—and they are always abstractions—can they fully or partially reflect the true sufferings of people in disaster even if they are based on personal testimonies? To answer this difficult question, one should first answer another one: "What was this conceptualization or model or whatever, developed for?" That is one of the crucial issues I tried to stress in my paper. If the answer to this second question supposes the description of sufferings or the like, my answer to the first question is negative. If the answer to the second question is some orientation for decision-making, my answer to the first question will be positive. While the process of decision-making involves individual persons, families, communities, and governments, it is the choice of a scholar to concentrate on any level. I chose the last one. Hewitt's general notion on this is right, but I do consider it a practical rather than a Western approach, because in Russia major decisions concerning disaster problems, supported by actual funding, are still concentrated at the top level. And Russia needs such major decisions no less than, of course, local ones.

Part II

FUTURE CONCEPTIONS OF DISASTERS

This section consists of eleven articles, in the main suggesting ideas about disasters that might be considered in the future. Dynes focuses on the need to locate disasters in a community context. The importance of the larger social order in thinking about disasters is discussed by Stallings. Rosenthal explicitly considers future social changes and how they could affect the way we should think of a disaster. Kroll-Smith and Gunter argue for the value of having a multiplicity of views on the central question of what is a disaster. How an anthropological perspective with an emphasis on culture affects the view of disaster is elaborated by Oliver-Smith.

Perry follows with a presentation of his reaction article to the previous five and particularly notes what he considers the development of a theoretical superstructure for disaster research. Each of the original authors then indicates what they think about what Perry has said about their contributions.

11

COMING TO TERMS WITH COMMUNITY DISASTER[1]

Russell R. Dynes

Concepts are most useful when their formulation leads to researchable questions. For social scientists, the concept of disaster needs to be rooted in some social unit—the choice here is the *community*, a universal form of social life and response. Since disasters are normatively defined and are manifest by extraordinary effort on the part of community members, the most accurate indicators of disaster effects is found in the action and adaptation of community organizations. Two major community categories are identified—*autonomous* and *dependent*—while two non-community types—*sector* and *noninstitutionalized*—are also suggested. These different categories lead to different research paths. In any case, disaster as social disruption has to be viewed in a social-system context.

INTRODUCTION

Disasters create many difficulties, even for social scientists. Social scientists have to deal with concepts that also have popular meaning, and some of those meanings can carry with them high emotional content. When events invoke moral and emotional reactions, conceptual discussions about them can often evoke charges of moral insensitivity and professional arrogance. Also, interest in disasters cuts across disciplinary lines so one's own interest is given highest priority, while the interests of others are considered marginal or perhaps even trivial. Given those difficulties, it is often prudent to ignore or at least to downplay conceptual discussions. Periodically, however, it can be important to raise such issues. One should, however, in the discussion about disaster, disclaim responsibility to catalogue every sin, every trauma, all evil that is intertwined with human history. There is a more delimited mandate to be explored here.

Background

As a preliminary step, it is not necessary to do an exhaustive history of the concept of disaster. Two sources will suffice for our more limited purpose. The first is found in *Organized Behavior in Disaster* (Dynes 1974), which was both a description of the early work of the Disaster Research Center (DRC) and a review of prior disaster research. There, I note that there were four common usages of "disaster"—as agent description, as physical damage, as social disruption, and as negative evaluation. That review noted that, in many discussions, different meanings could occur interchangeably within the same sentence but, for the purpose of DRC's research program, the most central meaning was social disruption.

While there have been periodic attempts to deal with the concept, a second source, a 1995 special issue of the *International Journal of Mass Emergencies and Disasters*, deals more directly with conceptualization. Quarantelli (1995e) posed the question "What is a disaster?" to five researchers with different social science backgrounds. Their collective contributions were then critiqued and the authors were then able to respond. All of the authors focused in one way or another on the idea of disaster as social disruption. Gilbert (1995) offered three paradigms—the pattern of war approach, social vulnerability, and uncertainty, while Dombrowsky (1995a) emphasized disaster as the collapse of cultural protection. (I was also struck by Dombrowsky's use of the Promethean analogy to suggest that we often trim the solution to fit the problem, perhaps anticipating my discussion here.) Kreps' (1995a) emphasis on disaster as a systemic event and as a social catalyst fits well with my own intellectual outlook. I found Porfiriev's (1995) extension of the concept to a social geographical area to be an interesting extension prompted by a set of pragmatic problems he confronted in his research efforts. Horlick-Jones' (1995) ideas of disaster as symbolic events pointing to the loss of control in the modern world fits well with the others. In spite of their differing national and disciplinary backgrounds, the authors seem to agree that disasters are social in origin; that agent determinism should be avoided; that social disruption should be the focus; that the concept is socially constructed and, to a certain extent, there was agreement that the focus of research should be on the reacting and responding social organization. These agreements, however, provided ample targets for the reviewing critique.

In that critique, Hewitt (1995), after concentrating on selected aspects of the other papers, moved on to what he chose to call "exclusions" in the field—what the authors did not talk about. Gaps might have been a better descriptor since "exclusions" becomes pejorative. Hewitt suggests the authors are simultaneously theoretically naive and morally insensitive, since they did not attempt to account for either Hiroshima or Auschwitz. I would second the reaction of Dombrowsky (1995b) and the response of Kreps (1995b) to Hewitt's treatment of their discussions.

While disclaiming here the possibility of extending the concept of disaster to account for all of the evil in the world, past and present, there still might be utility for the concept within the social sciences. Reference has already been made to the multiple meanings of the concept in *Organized Behavior in Disaster*. Other recollections from that source need to be made. The book was intended as first in a series of studies of various aspects of crises events. Its title was deliberately chosen to counter the usual notion of disasters as being identified by the disorganization of behavior. The focus of the book centered on the ways organizations function in crises events. That focus was intended, since, at that time, there was a corpus of research on disaster victims, but little on how forms of social organization dealt with disaster. In particular, DRC was interested in sudden impact events that would provide the maximum conditions for understanding the functioning of social systems. This emphasis on organizations in sudden onset events led to a primary concern for the emergency period and subsequently has been the base for an argument that this stream of research was insensitive to victims and to slow onset events. The book, however, pointed out that "While large scale community disasters provide a kind of maximum test of organizational functioning, other lesser crises and stress situations can also provide the opportunity to obtain useful and basic theoretical knowledge about the operation of social systems" (Dynes 1974: 4).

In any case, the book recorded the evolution of a series of decisions that guided DRC research, rather than a comprehensive attempt to "define" disaster. At that time, we were interested in disastrous events as a source of organizational stress, since these events had the effect of increasing organizational demands and decreasing organizational capabilities. This dialectic of a demand/capability ratio was a useful heuristic tool through which to think about the social consequences of disasters. While that approach was more fully explicated in *Complex Organizations: A Sociological Perspective* (Haas and Drabek 1973, especially Chapter 7), it never served as a major point of departure for organizational theory or disaster research. The lessons derived from that experience were that for social scientists to study disasters, one had to start with a particular social location—a social unit. This identification of the social unit meant that familiar social processes could be utilized as explanatory variables. Too, existing theoretical schemes and methodological approaches could also be utilized. In other words, while the events might be extraordinary, the research approaches necessary would be usual and ordinary.

On the meaning of social disruption

The notion of social disruption, as disaster, is more difficult. The traditional solution to this problem has been to infer social disruption directly from physical damage. The general logic is that the higher the number dead and injured, and the greater the property damage, the greater the social disruption. This logic has considerable appeal, especially when the physical sciences and engineering disciplines use quantitative measures to indicate physical disruption. Eight points on the Richter scale is worse than six or seven. In a hurricane, wind speeds of 200 miles an hour are worse than 100-mile-an-hour winds. And a hundred dead is worse than ten dead.

In multiple measures, however, you have considerable problems with combinations. Is ten deaths worth a hundred injured or worth a million dollars in property damage? Obviously such measures have to be evaluated in a specific social context. Ten dead in a community of 200 is more disruptive than in a community of one million. A million-dollar property damage to one family is different from that aggregate sum in a community of 10,000. Also, the death of an 80-year-old is less likely to be socially disruptive than someone at thirty. So, not all deaths can count as one.

One possible solution to the multiple measures is to talk about thresholds. Kreps (1995a) alludes to this issue in his recent review. If one establishes levels of damage, when these levels are exceeded then, by definition, a disaster has occurred. While this has the appeal of some standardization, such thresholds are of little value for comparative studies in different social systems, and economic thresholds quickly become eroded by inflation. More important, the mental gymnastics required to move from physical damage to social disruption are of Olympian proportion. It is easy to argue that governments should collect *real* social indicators of disruption, but it is unlikely that such an innovation will become a reality. The collection of social statistics is based on the identification of *stable* social characteristics, and administrative units are never coterminous with an impact zone. Too, post-impact collection of social disruption data would take time. While this might reassure some researchers about conceptual comparability, it is not likely to convince those who have experienced the disaster about the validity of those retrospective measures.

While we wait for improved measures of social disruption and their speedy post-event collection, which would lead us closer to conceptual nirvana, we need to find other ways to entertain ourselves. If you are going to study disaster as social disruption, you first need to identify the social unit. That means that you are going to have to develop more reliable social/behavioral measures, not measures of physical damage which guarantee post hoc explanations. If one shifts from aggregate indices of harm to focus on measures of nontraditional efforts of those in the social unit, then "disaster" will make more sociological sense. It should be clear that this approach is not intended to explain unique events in human history, such as Hiroshima or Auschwitz, but only those repetitive experiences in human communities in different social and cultural contexts. We now move, then, to a more sociological view of disaster and to indicate areas of research which emerge from that conceptualization.

TOWARD A SOCIOLOGICAL CONCEPTUALIZATION OF DISASTER

Let us start with an initial conceptualization which has considerable value for the following reasons:

1 It is based on a social unit.
2 It is based on a social unit that has cross-national and cross-cultural applicability.
3 It is a social unit that has the capacity and resources to activate a response to the disaster.

The particular social unit—the *community*—is a universal focus of social activity. Every community occupies physical space and has, in most cases, territorial boundaries so that the social entity can be characterized in part by its terrain and climate conditions. Communities have names and some degree of permanent settlement. But these physical, legal, and material features are only one dimension since communities are very complex systems of human activity. It is useful to think of a community as a structure which has evolved to meet needs and to deal with problems as well as to allocate resources to problems. This allocation process takes place within an organized division of labor as groups and organizations engage in efforts relating to one or more community need. Thus, the community has to be conceptualized as a multiorganizational system. In this conceptualization, the location of social action is the community.

So with the focus on the community, it is clear that disaster is defined in the emergency period. As we have already suggested, disaster agents are not self-evident. History is replete with examples of how communities are able to justify effects by religious and political ideology. This underscores the notion that disasters are socially constructed. (For an excellent treatment of the social construction of earthquake threat at the national level, see Stallings 1995.) The following formulation would seem to capture the relativity of the concept.

> *A disaster is a normatively defined occasion in a community when extraordinary efforts are taken to protect and benefit some social resource whose existence is perceived as threatened.*

There are several implications of that type of formulation. It is usual to talk about disaster *agents* creating disaster *events*. In our formulation there is no reference to disaster agents which implies that all disasters are socially caused. Specific names or traditional distinctions, such as God/Man, Natural/Technological, are less statements of causation than they are remnants of yesterday's normative arguments. Also, yesterday's inattention may be tomorrow's disaster and a disaster in one sociocultural system may not be in another. In addition, for the term "event" it is better to substitute "occasion". Event can imply a determinism that is not intended, a predetermined outcome. Occasion suggests more effectively the notion of an opportunity for something to happen.

The next question centers on identifying the norms of definition and the sources of definition of social harm. Quarantelli (1985b) has suggested a number of dimensions which may be central to evaluating social harm. These would include the proportion of the population in a community which is involved, the social centrality of the involved population, the length of involvement, the rapidity and predictability of involvement, the unfamiliarity of the crises, the depth of involvement and possible recurrence. This would suggest that occasions would be defined as disasters where there was extensive damage to community resources and to the health and social status of those who are central to the life of that community (e.g., community leaders) and to those who are dependent on those community resources (e.g., children, older persons, the injured and infirm.) If such communities were involved rapidly and unpredictably, and if that involvement were expected to continue and perhaps to recur, it would be quite likely that such occasions would be defined as disasters.

The issue is more complicated, however, in the contemporary world. Rather than the evaluation process being limited to community residents and perhaps to political leaders who have interest and responsibility, in the contemporary world, the important mediating element in the evaluation process is the *mass media*. Implicitly, one of its major functions has been to define "disasters." While much of the research on the media in disasters has focused on the accuracy of the coverage, that emphasis perhaps misses the point. It may be more important to view media coverage in terms of transmitting symbols that prompt concern and stimulate citizen involvement. The fact of the persistence of disaster myths in media coverage suggests that the themes drawn from normative criteria—on damage, on populations that need protection, on the interruption of hope—play an important role in defining the situation for others.

However, focusing on normative criteria embedded in public opinion and in media coverage explains only part of the definitional process. Values need to be embedded in concrete social structures to influence action and activities. While one of the characteristics of the emergency period is the search for information, most of the *factual* information is not known until much later in the social process. In effect, response *precedes* the compilation of accurate information. This suggests that changes in behavior and social structure provide a more accurate measure of "damage" than do conventional physical measures. Suggestions for the measurement of those changes follow.

Organizational involvement as the key determinant

Analytically, the effort should be directed toward identifying behavior which reflects *extraordinary effort*, that is, elements which have their roots in the predisaster community, but which are refashioned. Several possible indicators are suggested. One would be the involvement of community members in behavior that is not mandated by predisaster roles. These are usually considered volunteers who participate through ephemeral roles (Zurcher 1968). While the participation of volunteers has long been observed in disaster research, it is seldom studied directly. There is scattered research that provides insight into rates and patterns of involvement. (For a more extended treatment, see Dynes 1994a and 1994b.)

Probably the most important summary measure of a community's extraordinary effort is reflected in what has come to be known as the DRC typology (see Figure 11.1) which conceptualizes the involvement of community organizations in disaster tasks and looks at the relationship of predisaster tasks and structure to post-"impact" involvement. Type I organizations carry on the same tasks with the same structure but often expand their conventional efforts by extending the workday and double-shifting. Type II organizations expand their structures to carry out anticipated disaster tasks. These organizations anticipate the involvement and use of volunteers and thus expand to cope with the extraordinary effort. Type III organizations have no anticipated emergency responsibility, but may become involved once they possess manpower and other resources. This might describe construction companies that become involved in debris clearance or even search and rescue. Type IV organizations have no predisaster existence, but become involved with new tasks and develop a structure to deal with those tasks.

The purpose of introducing the typology here is to suggest that the pattern of organizational involvement is the best indicator of extraordinary effort and a behavioral reflection of normative judgments in defining a disaster. Put as a more circular way, when emergency organizations become involved, it is defined as an emergency. Organizational action implies that normative criteria are being evoked. This behavioral indicator is a much

Organizational Structure	**Tasks:** Routine	**Tasks:** Nonroutine
Same as Predisaster	Type 1 Established	Type III Extending
New	Type II Expanding	Type IV Emergent

Figure 11.1 Organizational adaptions in crisis situations

more concrete evidence of the definition of disaster than abstract public opinion or retrospective assessment of physical damage.

While there have been several extensions and refinements of the typology (Brouilette and Quarantelli 1971; Weller and Quarantelli 1973; Bardo 1978; Forrest 1978; Stallings 1978; Dynes and Quarantelli 1980; Drabek 1987), it is useful to note here the work of Gary Kreps and his colleagues (Kreps 1989b; Kreps and Bosworth 1993). In exploring patterns of stability and change of disaster involvement, Kreps merged organizational and collective behavior perspectives and makes a distinction of four structural elements that are individually necessary and collectively sufficient for organization to exist. From four structural elements—domains, tasks, resources, and activities—Kreps develops a taxonomy of 64 structural forms. Starting with the pattern of D-T-R-A or formal organizing at one end (Established Organizations in the DRC typology) and A-R-T-D or collective behavior at the other end (Emergent Organization in the DRC typology). While allowing an analysis of greater complexity, Kreps' work provides support for both the existence of and the importance of the fourfold typology. For the purposes here of developing a taxonomy of community disasters, the fourfold typology will be sufficient to serve as the basis for the exploration of different types of community disasters. The purpose of the following distinctions is to point to different research questions central to different types. In addition, since "community" is a universal social unit, the discussion attempts to make the various types relevant cross-culturally.

CATEGORIES OF COMMUNITY DISASTERS

From the viewpoint of the community system, it is possible to identify several model types of disaster. The first type and the basic model is called the *Autonomous Community Disaster*. This type would fit many disasters in developed countries. The community system is the location of the impact and the response by local community organizations. That involvement reflects a consensus that extraordinary efforts are being undertaken to deal with the social resources which are being threatened. An important subtype of Autonomous Community Disasters is what will be called *Community Accident*. The difference implied here is that the response is focused on the activities of institutionalized emergency (Type I) organizations. In effect, it is a delimited disaster and better characterized in *accident* terms.

The second major type is what will be called *Dependent Community Disasters* which implies that additional response resources are provided by other social systems, external to the community. Three subtypes are identified: (1) *Conflict Dependent*, (2) *Client Dependent*, and (3) *Proxy Dependent*. These are all situations in which the local community is seen as dependent by both national and international external agencies, which may become involved. This, in effect, creates a "dual" system, which creates an emergent pattern of organizational involvement.

There is a final category added for completeness and that is what will be called non-community disasters. The first subtype is called here a *sector/network* disaster and the final subtype is labeled a *noninstitutionalized* disaster (which is an oxymoron). In effect, these two subtypes represent conditions where there is limited consensus on social harm as well as limited institutionalization within community organizations about the nature and propriety of involvement (see Figure 11.2).

The rationale for the development of different disaster types is not to create meaningless and academic distinctions, but as a basis for illustrating important similarities and differences among types. One of the persistent problems of the interpretation of research has been that conclusions are drawn based on one disaster type and then generalized to other quite different types. The rationale here for the taxonomy is to point to different research questions.

The major difference among the types is centered in the notion of the capability of communities to respond on the basis of their own social resources. Resources here are conceptualized in terms of the organizational structure of the community. I am assuming that there will also be considerable complexity of informal activity. This, Barton (1969) has called the mass assault, which is helping activity on the part of persons, small informal groups, and families, and which would constitute an important part of the total community response. The more formally organized structures of the community, however, constitute the core of the organized response.

Autonomous community disasters

Two subtypes are differentiated: (1) community accidents and (2) community disasters.

Community accidents

These are situations in which an occasion can be handled by Type I or emergency organizations. The demands that are made on the community are within the scope of domain responsibility of the usual emergency organizations such as police, fire, medical and health personnel. Such accidents create needs (and damage) which are limited to the accident scene and so few other community facilities are damaged. Thus, the emergency response is delimited in both location and to the range of emergency activities. The primary burden of emergency response falls on those organizations that incorporate clearly deferred emergency responsibility into their domains. When the emergency tasks are completed, there are few vestiges of the accident or lasting effects on the community structure.

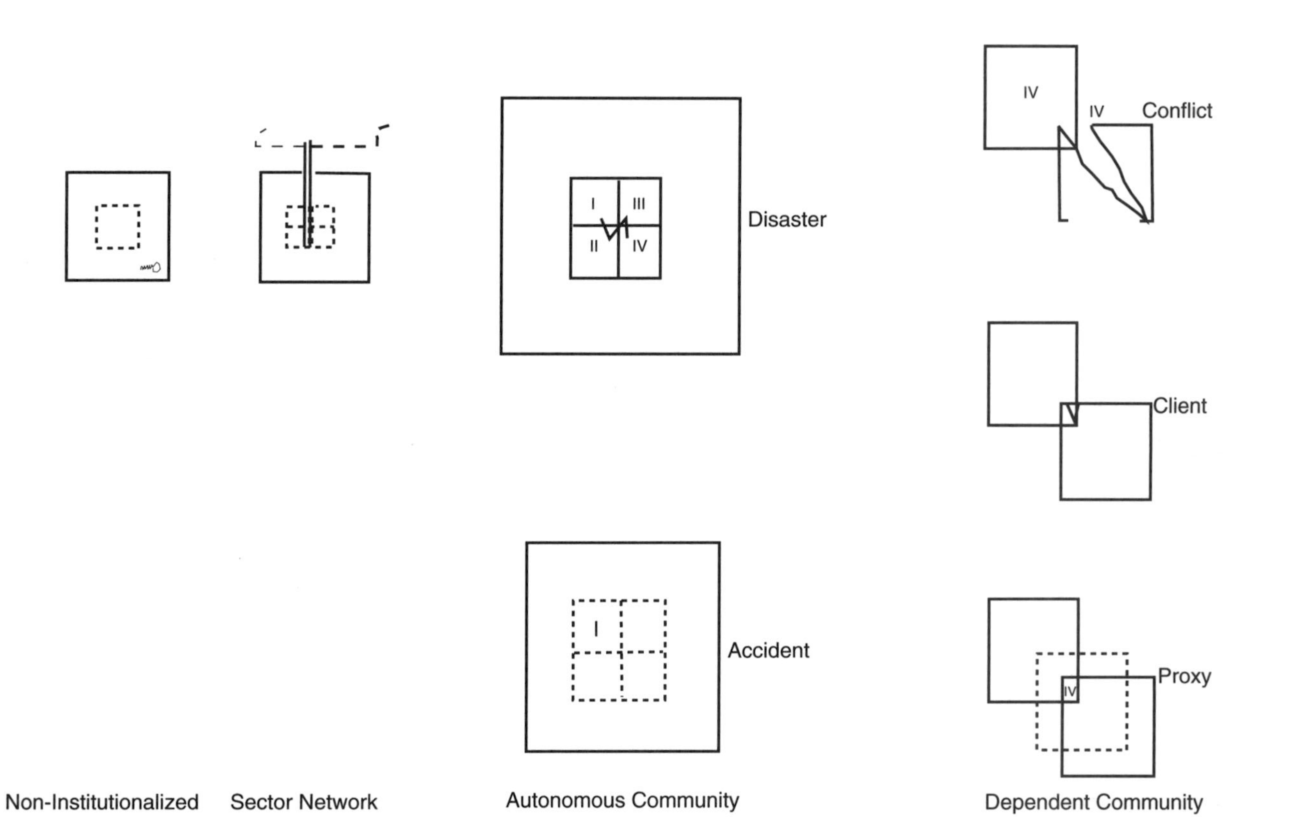

Figure 11.2 Types of community disasters

RESEARCH FOCUS

In these situations, research interests might focus on search and rescue, delivery of emergency medical services, security at the disaster site, coordination of multiple emergencies, handling of temporary interruption of community services, etc. Another focus could be on the "first responder," on the implementation of mutual aid pacts, the emergence of patterns of coordination, study of convergence on the accident site, and the social control of convergence.

Community disasters

This type represents the more traditional disaster. Differentiating this type from a community accident is the extensiveness of the involvement of organizations and other segments within the community. In community accidents, the emergency organizations will have developed some familiarity and accommodation to the domain definitions of other Type I organizations. In a community disaster, the pattern of damage may extend to several different places in the community rather than being focalized as it is within a community accident. Also, a number of community structures, perhaps including those that might house the traditional emergency organizations, might be damaged or destroyed. To determine whether such conditions exist requires the collection of information from other organizations. The increased involvement of other nonemergency organizations then creates the need for coordination of activity and for new patterns of communication among parts of the community that previously had no reason to communicate.

The need for coordination and the development of new forms and channels of communication has been termed *response-generated* demands as opposed to *agent-generated* demands. In other words, they are demands that arise because of the response itself and not because of the agent. (This distinction, however, is frequently overlooked during the emergency and is often ignored in disaster planning, which assumes that the demands being made on the community organizations derive from the disaster agent itself. The combination of *agent-generated* demands and *response-generated* demands creates a new and generally unfamiliar complexity to social relationships within the community.)

In terms of the previous comments about slow and gradual onset disasters, a sudden onset disaster would involve Type I and II organizations in rapid mobilization, quickly followed by Type III organizations and the rapid emergence of Type IV, while gradual onset would involve a more deliberate sequential pattern of I, then II, then III, and perhaps then IV organizations.

RESEARCH FOCUS

Many of these ideas are already reflected in the literature so that some of the research focus would be on the elaboration and replication of those notions. The time-phasing of organizational involvement has not to my knowledge been studied directly. Also, much more needs to be done on *response generated* demands. (Kreps and Bosworth (1993: 454) do confirm this pattern of involvement in footnote 10.) This category would encompass most disaster cases occurring in urban areas in developed countries and perhaps in most developing countries. It is important to note that the same physical agent, such as cyclones, might create several *different* disaster types within communities that are in close geographical proximity.

Dependent community disasters

In certain ways, these disaster types are extensions of the previous type, except that the local community response is compounded by outside assistance. This perhaps implies that in such situations the capacity of a community is "weak," incapable or perhaps even nonexistent. That may be the case, but in actual experience it would seem that higher levels of government as well as other extracommunity non-governmental agencies make a prior determination within their domains to provide assistance. That definition of obligation overrides and precludes determination of need. There may be examples of where community organizations are overwhelmed but usually that assessment is made by organizations external to the community as a matter of course in justifying its involvement. Such external involvement, of course, may be requested by local officials or at times perhaps by uninformed and inexperienced officials. In any case, the differentiation of this type from the previous type is marked by extensive organizational involvement of extra community organizations.

Three different dependent community disasters can be identified. In all of the subtypes, the assumption is made by organizations external to the community that the local response capacity is weak, damaged, or nonexistent. The three subtypes are: (1) conflict dependent, (2) client dependent, and (3) proxy dependent.

Conflict dependent

Perhaps a better term to use would be violent conflict or the concept of civil strife. Certainly, conflict is a common feature of every community. However, conflict usually operates within a context of some normative limits, e.g., within the governmental process. There are many occasions when violence or force, or threat of force, is used as a method of conflict directed toward some political end. This is an area of many complex issues in conceptualization that will be slighted here. However, the simple observation will be made that aspects of violence often become institutionalized to the extent that units external to the community see themselves as necessary to support the local deteriorating and perhaps polarized community organizations. Such external interests may serve to strengthen perceptions of unfairness and can lead to further divisiveness. Increasing divisiveness is then seen as justification for additional external assistance. The pattern of organizational involvement, by its very nature, is *emergent* and a frequent outcome is the creation of a dual assistance system, somewhat isolated from one another and at times opposed to one another.

Client dependent

A rather common pattern of disasters, especially in developing countries, is what can be called client dependent disaster. The assumption is made that the local community is unable or incapable of dealing with the range of disaster demands. Thus, high levels of government assume that such communities have to be supplemented or "strengthened." In certain instances, this assistance could be the result of disaster preplanning, but in most cases the judgment is made case by case, so that the pattern of organizational involvement is almost always *emergent*.

Proxy dependent

These disasters are defined most frequently by the media, and national and international organizations, relating to gradual and perhaps chronic demands that over time are assumed to have lowered the capacity of community systems to act as a responding unit. To a large extent, the response community is a surrogate composed of fragments of previous social structures. Those fragments may come from the consequences of other disasters when in response, smaller social units such as families, have migrated. The interest here, however, is not on tracing the complex casual links, but on the notion that, at some point in time, a "catchment" area develops and is identified as containing aggregates of people who have been earlier disenfranchised, i.e., hold citizenship in no viable community. These circumstances result in the creation of an "ad hoc" community or "surrogate" community, an amalgam of many cumulating local, national, and international elements of social structure. That process creates a new "community" with the primary function of responding to immediate disaster needs as well as to develop longer-term solutions, perhaps the reestablishment of some "real" community.

RESEARCH FOCUS

Certainly one common thread among the three subtypes is the emergent system that characterizes the disaster response, in large part because prior disaster planning is likely only possible by external agencies. Consequently, the pattern of response then centers around the needs of the external agencies, rather than of the clients. In effect, the emergent systems are likely to be rather paternalistic. Perhaps instances that do not fit the pattern of paternalism should be especially sought out to study.

In conflict dependent, the dual system might best be studied from the viewpoint of the community conflict literature and in terms of political and social movements. There is some literature about the differences in the functioning of emergency organizations in conflict and consensus disasters. There is also some literature on forms of "deviant" behavior in the contrasting situations as well as the emergence of new "accommodating" leadership roles. Not a great deal is known of the longer-term consequences of community violence and the adaptation which family units make to that, although considerable insight might be derived from wartime situations.

In the proxy community, a research focus could be directed toward the continuities of social life that persist among the victim population(s), continued patterns of migration, the reinfranchisement process, the integration of local and external elements in the social reconstruction of the community, differential patterns of response by different international agencies and by differing organizational philosophies, the shifting pattern of community needs in relationship to external political considerations, etc.

It would seem to be that most famine, drought, and perhaps refugee situations should be studied from the viewpoint of the *proxy dependent* community; at other times as *client dependent* and perhaps, on occasion, as *autonomous community* disasters. There is no reason to assume that they should be any different from any other agent in having differential effects. It is quite possible that a more detailed typology of proxy community could be developed by examining case studies. Such research might result in more complexity or perhaps the category does not reflect a core of reality. There is some literature on the creation of intentional communities and there is also a scattered literature on relocation and resettlement that provide certain hypotheses.

Non-community disasters

There are two other disaster "types" which will be mentioned here for completeness, although they may not merit extensive attention at this time. The important differentiation here is that the central analytical social unit is not the community. In these types, "impacts" and the response is focused on other social locations. The primary example will be identified here as the sector/network "disaster."

Sector/network "disaster"

This type is best explained by contrasts to the previous discussion. In contrast with community disasters, demands are primarily confined to one sector (institutional area) of the community and thus have little significance for the broad range of potential emergency organizations. That is, the effects do not directly affect normative domains of many other organizations. Thus, the "disaster" is a sector "problem" rather than a "community" problem. The response structure does not take the usual community format but is sectorial, linking a network of people and organizations together; however, that response does not demand extensive or total involvement of most community emergency organizations.

It would seem that most current environmental issues are best described as sector disasters, as well as responses to most disease entities. It is also important to note that there are parallel subtypes in reference to *sector accident*, which would now describe most incidents of hazardous materials spills in developed countries. In addition, there are examples of what might be called dependent sector disasters, which now characterize many of the environmental–ecological issues in developing countries. The network within the sector links persons and organizations within the local community with others at the national, and sometimes the international, level. Those linkages often create the opportunities for potential conflict when national and international members of the network demand greater local concern and involvement than the "locals" feel is merited.

Research on sectorial disasters would substitute a social network focus for research, rather than a community focus. In addition, the perception of community members as to the obviousness and seriousness of demands would be researchable. One might hypothesize that most community members would reflect rather little normative concern. Consequently, the network might attempt to create a heightened awareness of the disaster demands and their consequences. The vocabulary they use to define the problem may be apocalyptic and epidemic. In these types of disaster, the media might play important "defining" roles. In fact, one of the strategic directions of sector networks would be to convince other sectors in the community that sector disaster is actually "community" disaster. That status might describe Love Canal in relation to toxic materials as well as San Francisco in reference to AIDS. Careful research on the expansion of sector disasters to community disasters might provide one research focus.

Non-institutionalized disaster

The final "type" is a contradiction in terms of the theoretical scheme just presented, so an accurate label is difficult to find to convey empirical reality. Perhaps the terms of "near-disaster," "public opinion disaster," "movement disaster," or perhaps "non-disaster" might describe certain cases that are at the margins of consideration, especially when the term has been defined here in terms of the institutionalization and involvement of emergency organizations. Its very description of not being well institutionalized within organizations could preclude its consideration within the typology. There are periods when "potential" demands become a part of public discourse. That discussion centers on "shifting" what previously have been considered "personal" problems to the level of concern that deserves institutionalized attention within the community. Such public discussion centers on the criteria of social harm, the capability of the victims, their characteristics, and the scope of social responsibility. That is, there are discussions about effects, about victims, and social responsibility.

These issues are not only the focus of media attention, but can be a focal point in the development of social movements and political protests. It is quite possible that these "non-institutionalized" disasters are simply an early developmental stage of sector or community disaster. It may be the identification of social harm, particularly among innocent victims, is a necessary precondition to the discussion of the location of social responsibility. Too, the focus of social responsibility may shift from "private" to governmental organizations in that developmental process. While this is not the place to further explore these issues, they should offer many research opportunities. The issues, however, are more likely to utilize theories of mass communication, collective behavior, and social movements than organizational and community theory. A careful examination of historical materials might reveal a "stage" theory of disaster more clearly linked with their "origins" in social movements, rather than linked to physical conditions.

Other considerations

While the previous discussion has focused on the community as a key analytical unit, there are obviously other choices, other social units—the family, task sub-systems such as search and rescue, political and administrative systems as well as regional, national, and international systems. Wherever the starting point, it is important to keep system interrelationships at the core of research. Nowhere is this more important when the focus is on individuals. Individual responses always have to be interpreted in some social context, not as some inherent personality trait or as some inferred cultural trait, such as fatalism. As a general principle, behavior at lower-level systems can seldom be explained without understanding the social context in which that individual operates.

Also, while the previous discussion has focused on social structure, that focus centered on a specific time phase of the disaster occasion—the emergency period. By implication, that formulation implies the possibility of viewing the disaster occasion along some continuum of social time. In general, a common vocabulary has emerged which includes mitigation, preparedness, emergency response, and recovery. Those stages should not be measured in chronological time, but as a characterization of activities and processes. The various stages are intended to exhibit continuity and are, in effect, circular in nature. I would argue, however, that the emergency response phase is most important in understanding the entire process, since the combination of preexisting community organizations and emergent behavior are the critical elements that affect subsequent phases of the process.

Of course, there are other conceptual possibilities that allow certain topics to be understood with some degree of completeness. For example, it is useful to take a social systems approach in the consideration of topics, such as warning, since that process involves action by organizations that monitor threats, transferring information to organizations that prepare warning messages which communicate those messages to various "populations." Those populations interpret those messages in differential ways which evoke several forms of social interactions and, ultimately, a range of behavioral responses. This complex social process involves several stages, as well as several different levels of social structure. It can best be treated as a middle-range theory so that incomplete knowledge can more easily be identified. There could be other "middle"-range theories centering in such concepts as evacuation, relocation, mass assault convergence, organizational change, interorganizational coordination, and long-term community change.

Finally, there are always opportunities in disaster occasions to test theories and concepts derived from completely different contexts. For example, when the Disaster Research Center started organizational research, the initial models, drawn from the existing organizational literature, were found to be too static to deal with organizational behavior in disaster. Consequently, this led to the development of other conceptualizations, such as the typology of organizational involvement introduced earlier.

If more general theory has validity, then it should have validity in the disaster occasion. For example, family decision theory should work in the decision to evacuate and/or family adjustment theory should be applicable in understanding the recovery process at the family level. The point is that good theories of behavior should be applicable in disasters. If they are not, then they are not good theories.

CONCLUSION

It is necessary to conclude even when concepts are obviously open-ended. For this sociologist, disaster means observing the disruption of some social unit. One conceptualization of the disaster occasion was presented based on the idea that disasters are always normatively defined. The locus of such definitions has been the community, since that form of social life is always the primary responding unit. The indicator of extraordinary effort was organizational involvement, since disaster demands have to be related to the community's organizational capabilities. Two major disaster occasions—Autonomous Community and Dependent Community—and two non–community-based types—sector network and "non-institutionalized" disasters—have an identifiable base in the organizational structure of the community. Possible research leads are offered for the different types. Alternative conceptions of disaster occasions can also be based on social processes, systems theories and also can be used to test more general theories within the social and behavioral sciences.

NOTE

1 Many of the ideas in this paper were initially generated in discussions with Everett Ressler at the Asian Disaster Preparedness Center, AIT, in Bangkok. His experience in a variety of situations in Asia, Africa, and Latin America insured the discussions moved in a cross-cultural direction.

12

DISASTER AND THE THEORY OF SOCIAL ORDER

Robert A. Stallings

"What is a disaster?" is the wrong question, in my view. The basic question facing sociologists who study disasters should be: "What do we need to explain?" or "Toward a theory of what?" The answers offered in this chapter are that we should be trying to explain the nature of societies and that our primary task therefore is the continued development of a theory of social order. The empirical study of disasters is one—but only one—part of such a sociological project.

This chapter contains three major sections. The first section extracts several useful ideas from the five papers plus the reaction paper that were originally published in the November 1995 issue of the *International Journal of Mass Emergencies and Disasters* (IJMED). This section is highly selective and not intended to describe the complete contents of these papers. It is limited to a discussion of what seems most useful to the construction of a theory of social order, one that encompasses disasters.

The second section discusses three reasons why the construction of a sociological theory of disaster per se does not seem worth pursuing. One reason is that definitions of disaster will always include a non-social component. This occurs because of the way in which events are selected for study. Another reason is that a sociological definition of disaster at best identifies only a setting or context. Developing a theory of social settings, rather than of the processes that take place within them, seems misguided at best. A third reason is that a sociological theory of disaster would necessarily be a theory of the middle range. However, middle-range theories tend to be logical dead ends because their propositions are difficult to apply to other phenomena.

The third section outlines one direction that a program to develop a theory of social order might take. A conceptual model is introduced that describes social structure as an amalgamation of routines continually reenacted over time. When routines are disrupted, they may either be abandoned entirely or modified to incorporate the disruption, or some countermeasure may be invoked in an attempt to restore the routine. Exceptions is the name given to disruptions, interruptions, and breakdowns of routines; the countermeasures invoked to restore them will be called exception routines; and the interrelated routines, exceptions, and exception routines comprise the process of routinizing social structure. Existing work that already moves in this direction is identified.

A concluding section discusses what a research program would look like if it were focused on the empirical examination of existential definitions. It also identifies some of the risks that the proposals in this chapter may pose for the discipline. Before beginning, there is a set of terms that will be used frequently in the following discussion. These terms are useful in discussing the subject of definitions.

Sociologists are not alone in the use of constructs, concepts, and definitions. The people whom sociologists study also use constructs and definitions to select, organize, and make sense of the world around them. Schutz (1967: 5–6) distinguished what he called the common-sense constructs that laypersons use from scientific constructs. The latter are "constructs of the second degree, namely constructs of the constructs made by the actors on the social scene, whose behavior the scientist observes and tries to explain" (ibid.: 6). McKinney (1969: 2) relabeled Schutz's terms, calling them existential and constructed types. Existential types are "constructed by participants in social systems. They are fundamental data for the social scientist and stand in contrast to, and yet in continuity with, the 'second order constructs' or *constructed types* he develops and utilizes" (italics in the original). Modifying slightly previous terminology, the term disaster as used in the social sciences will be called a constructed definition. Definitions of disaster used by all other actors will be referred to as existential definitions.

THEMES FROM THE FIVE PREVIOUS PAPERS AND REACTION PAPER

We start with a basic idea that seems trivial, but is not. However they may be defined, disasters are significant events (Kreps 1995a; Porfiriev 1995). The disruption associated with disaster is, by customary standards, non-trivial. Disasters are neither confined to isolated subsystems (a single household) nor are they of fleeting duration. Kreps continues the tradition of defining disasters in terms of physical events (albeit socially defined) which become sociologically relevant because of their consequences (they are social catalysts). The locations known as disaster areas are clearly recognizable to all as containing physical destruction that disrupts the routines of everyday life. Both Kreps and Porfiriev insist that "true" disasters are societal (i.e., macro-level) phenomena. That is, their significance extends beyond the disaster area itself. Kreps keeps the focus at the macro-level on events that affect "societies or their larger subsystems" (1995: 258).

Disasters involve the disruption of important societal routines. Gilbert's description of vulnerability and uncertainty paradigms equates disaster with disruption of the fabric of society. Describing the vulnerability paradigm, he writes that disaster "is the result of the upsetting of human relationships" and "is virtually experienced as a process whereby specific area activities carried out by the actors and the structures of the community, begin melting" (1995: 235). Describing the uncertainty paradigm, he argues that uncertainty is a correlate of multiple symbol systems, not the product of an absence of information but the result of "the anarchical profusion of information" (ibid.: 237). Thus, disaster "means the loss of key standpoints in common sense, and the difficulty of understanding reality through ordinary mental frameworks" (ibid.: 238).

Dombrowsky, following theoretical leads laid down not only by Carr (1932) but also by Clausen (1983, 1992), goes even further. He argues that not only the consequences but also the causes of disaster are connected to social structure. He points out that the majority of existing definitions of disaster, scholarly and non-scholarly, let "society" off the hook by "falsely" explaining disasters as caused by "overwhelming forces coming from outside" (Dombrowsky 1995a: 246). He observes that these external forces fail to produce the disruption labeled disaster whenever societies have the means to protect themselves. (Much of his argument is based on a comparison of the differing effects of heavy snowfall on southern Denmark and northern Germany in 1978–1979; see Dombrowsky 1983). He says that since disasters are therefore products of society's "lack of capacity" to defend itself from external forces, is it not logical to ask: "Who is responsible for such a lacking?" (Dombrowsky 1995a: 246).

For a sociological answer, he revisits the early article by Lowell Juilliard Carr (1932). If damage could be prevented or reduced through human protective action, then disaster—the physical consequence of the intersection of society and natural forces—would not exist. Disaster is a function of knowledge, in other words (Dombrowsky 1993). When knowledge is adequate, no external force can produce disaster; ships ride out storms, buildings shake but do not collapse in earthquakes, flood levees hold, etc. (Dombrowsky 1993: 247; see also Carr 1932: 211). When knowledge is inadequate, disaster results. Thus, following Carr, Dombrowsky defines disaster as the collapse of cultural protections. That is, habits, folkways, beliefs, laws, etc. either deflect or they fail to deflect the natural forces to which societies are exposed. Failure of these cultural protections results in the disruption of routines labeled disaster. Dombrowsky's paper suggests the need to examine ways in which social routines invite their own disruption.

Increasing attention is given in these papers to the relationship between culture and disaster. Dombrowsky emphasizes culture as a defense mechanism and disaster as a failure of that mechanism. Gilbert cites a disconnection between the disruption of disaster and the everyday meaning provided by the world taken-for-granted. Horlick-Jones (1995) is even more explicit. Citing some of the work by Anthony Giddens on postmodernism (Giddens 1990, 1991), he argues that disasters pose a threat of destabilization. Implicitly introducing the hypothesis that belief in order is an essential ingredient of order, he argues that the certainty and predictability required for everyday existence in postmodern societies involves "trust in a range of institutions served by anonymous individuals" (Horlick-Jones 1995: 311). Such institutions and the individuals who manage them are "charged with controlling risk" (ibid.: 311). Accidents and disasters contribute to "[a] widespread sense of betrayal by public institutions and doubt in their ability to manage risk" (ibid.: 312). The potential for a crisis of legitimization calls attention to the need to understand the role of all institutions, including the state, in the process of (re-)legitimization following the disruption of social routines.

Porfiriev also is explicit about the challenge that massive disruption associated with disaster poses. He, too, argues that disasters threaten social stability. He defines disaster as "a state/condition destabilizing the social system that manifests itself in a malfunctioning or disruption of connections and communications between its elements or social units . . . making it necessary to take *extraordinary or emergency countermeasures to reestablish stability*" (1995: 291; italics added). More often than not, the threat of instability is sufficient for the state to intervene. State intervention in disasters to reestablish "public order and safety" is a point also emphasized by Dombrowsky (1995a: 242–243; see especially Dombrowsky 1989: 103–129). The role of the state in defining and reacting to exceptions that it sees as potentially destabilizing is central to a theory of social order.

Hewitt's reaction paper, in a sense, lays out his own theory of disaster and social order, albeit a radical populist version of one. He proposes that disasters are products of internal societal forces rather than external agents. He explicitly takes Horlick-Jones arguments about the relationship between laypersons and institutions farther, seeing in this relationship the cause of disasters. Modern institutions, in particular the state, have ever-increasing control over everyday affairs. This, in Hewitt's view, increases the risk of disaster since the experts who run these institutions are not up to the task while, at the same time, the people who are exposed to risk have less and less control over the direction of public affairs (1995: 325). It is the lack of accountability of the institutions of the modern state that causes disasters. According to Hewitt's theory, disaster reduction will occur only when people take back power from the state, exercising grass-roots control over their own fate—and over the social order (Hewitt 1995: 332–333). Whatever its causal role, the state is clearly central to the examination of routines, exceptions, and exception routines.

One final point emerges from the six original papers. Dombrowsky is most explicit about it. He reminds us that *sociology is not the only source of important definitions of disaster* (1995a: 242–243). The groups and organizations that we study have their own definitions. Dombrowsky illustrates with three examples (the German Red Cross, German federal law, and German insurance companies; his list could easily be extended by including other definitions such as those found among laypersons, news organizations, hospitals, politicians, etc.; see below). Horlick-Jones agrees and argues that non-academic definitions of disaster are a function of organizational interests; that is, they "strongly reflect the preoccupations and circumstances of whoever is doing the [defining]" (1995a: 307).

Dombrowsky goes on to suggest that, in definitions of disaster, "[t]hose who define declare what they intend to do with the social processes called disaster" (1995a: 242). Existential definitions always emphasize the physical (non-social) properties of events because these are the "triggers" that determine the invocation of countermeasures. The result is that over time organizations define disasters "according to the solution at hand," thus "cutting reality into the parts that fit the organizational capabilities to handle them" (ibid.: 244) and ignoring the parts that do not fit. "Reality then is exclusively seen from one approach; the solution defines the problem, and deductively, reality. 'Disasters' are predominantly defined in this way" (ibid.: 244).

In sum, the following ideas can usefully be extracted from these six original papers. Each idea stands out because of its importance to a theory of social order. Disasters affect entire societies; they are neither trivial nor confined to localized social units. Disasters involve the disruption of everyday routines to the extent that stability is threatened without remedial action. Increasingly significant is the loss of certainty and the undermining of faith in orderliness. The state is a major institution for supplying countermeasures when routines are disrupted. The potential for disruption is contained within all social routines; each is vulnerable to breakdown. Existential definitions classify the various disruptions and signal the appropriate countermeasures to be taken.

THREE ARGUMENTS FOR RETURNING TO GENERAL SOCIOLOGICAL THEORY

In the introduction I suggested that the fundamental question sociologists who study disaster should address is: "What do we need to explain?" In this section I argue that a purely sociological (constructed) definition of disaster is unattainable. Disaster research depends on physical (non-social) properties such as type of agent, extent of damage, and amount of loss. Furthermore, the current state of affairs will not be improved by pursuing a theory of disaster. Disaster is a setting or context for social processes, not the processes themselves. A theory of disaster is thus a theory without content. Finally, pursuit of a theory (or, more accurately, theories) of processes in disaster is likely to produce still more middle-range theories. However, middle-range theories quickly become logical dead ends because they cannot be applied to study processes outside the disaster context. Therefore, a return to general sociological theory with its central concern for sustaining order in the face of uncertainty—including uncertainties created by the destructive effects of disasters—seems like a viable alternative for improving the quality of our work.

The inevitable non-social component of disaster definitions

Regardless of how one formally defines the term disaster, in practice the definition will always have a physical component. The physical properties of events are triggers for disaster researchers, much the same way as Dombrowsky (1995a: 242–244) describes them as triggers for organizational countermeasures.

In order to study whatever disaster-situated process one is interested in, it is first necessary to know where and when that process can be located. In studying historical disasters, for example, this means consulting archives for records of "earthquakes," "floods," etc. For gathering primary data, this usually means waiting for word that an event, appropriate for study, has occurred. That word usually comes first from the news media which provide information about what has happened (e.g., an earthquake), where the event has occurred (e.g., in a rural area northeast of Lima, Peru), and something of its severity (typically in terms of preliminary Richter-scale readings plus early estimates of the number of dead and injured). If it seems reasonable to assume that the event involves phenomena of interest, researchers can attempt to reach the scene to size up the situation and determine which individuals or organizations are key players. On the basis of this initial reconnaissance, researchers can subsequently draw samples of units from which data will be collected. (For an excellent description of this process by a veteran disaster researcher, see Quarantelli, 1997). The process is similar in other types of research such as those on predictions and warnings: researchers learn of the threat (stage one) and proceed to draw a sample within the threatened area (stage two) (for example, see Mileti and Fitzpatrick 1993 especially pp. 105–113).

Technically speaking, a multistage sampling process (Babbie 1973: 96–99) is being executed. The physical event is the "primary sampling unit" (ibid.: 194), or the unit sampled in the first stage. (The fact that there is rarely more than one primary sampling unit selected does not negate the description of this as a two-stage sampling process.) The "final sampling units" from which data will be collected may be individuals, groups, or larger social units depending upon the focus of the research. Unlike other applications of multistage sampling wherein probability sampling is utilized at each stage (ibid.: 213–224), the selection of the primary sampling unit (i.e., the disaster event) is made using what can only be described as purposive sampling (ibid.: 225), a type of non-probability sampling. Final sampling units may be selected either randomly or non-randomly; non-probability sampling (identifying "key informants," the use of "snowball" sampling, etc.) has until recently been more frequently employed at this stage than has probability sampling (cf. Marks and Fritz 1954: 6–16).

It is the *physical properties of events* that makes disaster research possible. There is no sampling frame (Babbie 1995: 194) which lists the names, addresses, and telephone numbers for disaster-situated processes. News reports of disaster become the sampling frame. They identify the existence of the event (the primary sampling unit) and its location (a clue to how one can locate the names, addresses, and telephone numbers of final sampling units). In other words, the physical properties of events will always be part of the definition of disaster because it is knowledge of the physical properties of events that makes empirical research possible. Disaster-situated processes are frequent, recurrent, and normal, but they are not ubiquitous. To study them requires a two-stage sampling process, the first stage of which is to select an event based on its physical characteristics. Put more technically, existential definitions in everyday use, especially the existential definitions used by news organizations and relief agencies, are indispensable in empirical research and are inevitably related to the constructed definitions of disaster used by sociologists.

Equating disaster with setting

The desire for consensus on a definition of disaster stems in large part from the need to know how to generalize from research results. Without some agreement on its analytical properties, it is difficult to know what to make of specific empirical findings. For example, does widespread looting during a massive power failure in a large city undermine the hypothesis about the absence of looting in natural disasters? The answer depends on whether power failures should or should not be subsumed under the definition of disaster.

The present angst (see Quarantelli 1995b) results from the growing belief that, despite several decades of research, knowledge about disasters does not seem to be accumulating in the manner of a "normal science" (see Kuhn 1970; see also Cole 1992). The key to knowledge accumulation is the construction of theory. Clearly, it is undesirable to invent "little" theories in order to generalize about disaster (e.g., a theory of behavior in hurricanes, a theory of response to earthquakes, etc.). Problems arise, however, in knowing how to generalize from studies of destruction produced by other types of agents (e.g., technological disasters) and by events that differ in terms of basic analytical properties (e.g., conflict events). The desire to incorporate findings from such events is illustrated by the number of terms offered over the years to encompass events that have less and less in common with natural disasters. Here is an unsystematic sampling of such terms, arranged in alphabetical rather than logical or chronological order: crises, environmental uncertainty, extreme environments, severe environmental distress, system stress, and unscheduled events. However, these terms imply solutions that are no more satisfactory than "little" theories (e.g., a theory of crises; a theory of environmental uncertainty; a theory of extreme environments; a theory of extreme environmental distress; a theory of system stress; and a theory of unscheduled events).

The basic problem is that these would be theories without content if they are literally theories of settings. It is the processes occurring within these settings that we should explain, not the settings themselves. From the beginning, disaster has been a means to an end—the study of sociologically interesting and practically important phenomena—rather than an end in itself (see also Dynes and Drabek 1994: 7). It was a natural laboratory for the study of basic social processes that could not morally, legally, or ethically be produced in controlled experiments: "Disasters provide a realistic laboratory for testing the integration, stamina, and recuperative power of large-scale social systems" (Fritz 1961: 654). Those who ran the first systematic disaster research project in the world, the National Opinion Research Center's disaster study project of 1950–54, were clear about the major problem with which they were concerned: "The maintenance or rapid re-establishment of effective social organization and morale" (Marks and Fritz 1954: 1).

In other words, pursuing a theory of disaster per se seems misguided. However, a theory of social organization or a theory of morale that can benefit from the study of disasters does make sense. Borrowing from Luhmann: "For sociology, the topic of risk [and, I am arguing, disaster as well] ought thus to be subsumed under a theory of modern society, and should be shaped by the conceptual apparatus thereof" (Luhmann 1993: 6).

The limitations of middle-range theory

One diagnosis for the current condition of disaster research in American sociology is that it is suffering from its own success. That is, as the special area known as the sociology of disasters has evolved (with panels at regional, national, and international sociological conferences, with specialty journals that provide outlets for its research products, with a second and now third generation of academic researchers, etc.), it has moved away from some of the core questions that originally intrigued its founders. For example, disasters caught the attention of sociologists trained in collective behavior at the

University of Chicago (e.g., Charles Fritz, Lewis Killian, E. L. Quarantelli, Anselm Strauss, and Ralph Turner). The opportunity to study how lines of individual behavior were fitted together under circumstances in which institutionalized frameworks seemingly had broken down must have been exciting to students interested in a theory whose contents included milling, circular reaction, panic, mass behavior, and the like (see Quarantelli 1994: 27–35).

What was once a means to an end is now an end in itself. There is now the sociology of disasters instead of simply sociology—to which studies conducted on disasters contribute. There are a lot of empirical generalizations (see Drabek 1986), but there is also less certainty now as to what they add up to. This is not a new situation. Over the years, theories other than collective behavior theory have been used to try to put disaster research findings into a more general framework. In the 1960s, for example, organizational theory was used to synthesize findings about organizations under stress (Haas and Drabek 1973: 237–263; Drabek and Haas 1969). More recently there have been calls for treating disasters as a type of social problem (Drabek 1989a; Kreps and Drabek 1996; but see also Stallings 1991, 1995). These and other efforts signal, I think, am implicit recognition of the logical limitation of a theory of disaster.

This limitation was recently made explicit by Peter Blau (1995) in a reflective piece on his exposure to sociology as a graduate student attending Columbia University under the GI bill. He describes how the collaboration between Paul Lazarsfeld (best known for his empirical contributions) and Robert Merton (best known as a theorist) both symbolized and fostered the interconnectedness of broad theoretical concerns and highly focused empirical research. In this context, Blau discusses Merton's own work in terms of Merton's call for "theories of the middle range." He points out that Merton himself never dealt with what most sociologists think of as theories of the middle range. Rather, he worked with broad questions of social life, but sought answers to them in the explanation of specific social forms and processes. Blau argues that it could not have been otherwise because middle-range theories are dead ends. They allow no further possibility for their own development because they limit the sociologist to one subject (juvenile gangs, nurses, etc.).

> Genuine theory cannot be middle range, for it must be distinct from research in two fundamental respects: Its concepts must be abstract and its propositions general, lest it be not testable since it has no new or different implications. An exploration of gang warfare or of some criminal behavior . . . that deals only with that specific crime cannot be tested by making predictions for other conflicts or controversies. Indeed, Merton's middle-range theories are really not middle range. [Here Blau uses as an example the

> analysis by Merton and Kitt (1950) of data gathered from soldiers during World War II.]
>
> The theoretical explanation advanced in terms of reference group comparisons, however, does not refer to some differences between military police and air corps, but to those between any groups that differ in the chances of obtaining some rewards. This is not the middle ground between research findings and theory but a true theoretical generalization, albeit one derived from research findings, yet testable with new empirical predictions.
>
> (Blau 1995: 5)

The sociology of disasters seems boxed in precisely in the way that Blau describes. It is trapped somewhere between its empirical findings on the one hand and theoretical concerns of the discipline on the other. Having ceased to be a means to an end (the development of general sociological theory), the sociology of disaster is littered with theories of the middle range. There are theories about how organizations adapt, about how individuals process warnings, about how communities recover, and so forth. These are "stand alone" theories. Integrating them with general sociological theory has proven difficult.

Where should we go from here? Any definition of disaster will inevitably have a non-social component; a theory of disaster as setting does not seem to be about anything; and theories of processes found only in the aftermath of disastrous disruptions seem doomed to the limitation of middle-range theories. The most logical alternative is to return to a higher level of abstraction and employ a general sociological theory to study disasters and theoretically similar phenomena. I think that this is the only sociologically justifiable alternative and will try to outline its implications and its potential in the next section.

ROUTINES, EXCEPTIONS, AND EXCEPTION ROUTINES

Sociologists who study disasters should return to a focus on broad processes of social order. That is, we should turn our attention to the fundamental questions that challenged Durkheim, Marx, and Weber in the last century and that remain fundamental at the end of this one. These are questions about the basic processes that make societies possible, about how continuity is achieved in a social universe characterized by continual change, and about how order is accomplished in the face of uncertainty. Since I take conflict, change, and uncertainty as starting points, I therefore describe this effort as contributing to a sociological theory of social order.

To do this, I recommend that we continue doing much the same sort of empirical work that we have been doing for half a century, namely, studying disasters and such other crises as present themselves. What I think we must do differently is to think through much more carefully and systematically—both before we write proposals for funding and after we have gathered our data—what each specific inquiry contributes to a theory of social order. In my judgment, it is the big picture that is missing in too much of our work. That big picture would portray disasters as only one type of "interruption" that all societies in every epoch confront. However, I do not think that shifting to comparative and historical research alone is sufficient. A shift in the type of theory used to guide disaster research is even more important.

We should think in larger terms than what types of events constitute a disaster. Instead, we should ask: "Of what more general category is disaster only one type?" I would like to sketch one answer to this question. Disasters are fundamentally disruptions of routines. However, in order to avoid as far as possible the re-equilibrating baggage of terms such as "disruption," "interruption," and "breakdown," I prefer to call these exceptions. That is, disasters are just one among many types of exceptions that one can observe occurring within a social system. We know them to be exceptions because participants in the system designate them as such, thus distinguishing them from routines. (Existential definitions are employed in making this distinction.) Routines are the actions (e.g., getting ready for work) and interactions (e.g., meeting with one's clients) that are repeated over specific units of time (e.g., the workday). Actions and interactions repeated routinely provide structure for individuals' lives and, in the aggregate, constitute the structure of social systems. An exception is the breakdown, interruption, or disruption of a routine.

Exceptions are as natural as the routines for which they constitute disruptions. Put differently, no social routine continues indefinitely without experiencing disruption (e.g., sooner or later a college class will not be held because the instructor cannot be present). What is sociologically important is to identify and to study those routines whose disruption is deemed serious enough that mechanisms have been worked out to address them. Borrowing from the analytical strategy of Durkheim (1964 especially Book One: 49–229), the existence of mechanisms for addressing exceptions tells us something about the importance of the routines they are intended to restore, hence about processes of social order. I call such mechanisms exception routines to indicate that they themselves are routines, and I use the verb "to address" as a way of indicating that the existence of exception routines in no way means that they are effective or even appropriate in a given situation. In fact, exception routines may persist for reasons that have nothing to do with the reasons they developed.

For a variety of reasons, I am especially interested in those exception routines controlled by the state. More generally, my approach to elaborating a theory of social order focuses on state–society relationships. Some examples include the following: work as routine for an individual worker, unemployment due to downsizing by company management as exception, and government unemployment compensation, job placement services, and training programs as exception routines (this example of employment–unemployment–unemployment program nicely illustrates that for a given social unit—a family, a segment of the population of a city, a segment of the labor force, etc.—there is no reason that the exception [unemployment] cannot become routine [jobs that disappear permanently]); marriage as routine for a given couple, separation as exception, and divorce laws and procedures as exception routine; health as routine for most individuals of a certain age, illness as exception, and organized medicine as exception routine; and finally, traveling over a bridge as routine, that same bridge washed out by flood waters as exception, and the response to the flood disaster by public agencies as exception routine.

The point is not that unemployment, divorce, and illness—either in individual cases or in the aggregate—are disasters like floods. The point is that floods create exceptions to routines as do downsizing, domestic discord, and disease. Exception routines have been developed over time to address each. Why? What does the exception routine tell us about the routine? (It may tell us little or nothing about the exception.) For what routines does the state control exception routines? Why do exception routines exist for some exceptions and not others? Why are some exception routines highly centralized while others are decentralized (e.g., left in the hands of state or local governments)? Which exception routines are addressed primarily at prevention (e.g., creating requirements for marriage that discourage marriages likely to fail or at least creating requirements so that divorce is difficult to obtain) and which are aimed at the aftermath of exception (e.g., creating rules about the division of assets after divorce)?

What is an exception and what is routine may differ when viewed from the perspective of different social units. That is, exceptions within one unit may be routine for another. One's house does not catch on fire every day, but the local fire department routinely responds to residential fires. A given couple does not divorce repeatedly, but marriage counselors routinely meet with couples on the verge of divorce. A given community does not experience major disaster routinely, but FEMA (the US Federal Emergency Management Agency) personnel are repeatedly deployed in the United States every year. In other words, exceptions are simultaneously routines. Two related propositions follow from this. One, suggested by Marxist sociologist Wieland Jager (1977; for a brief English statement, see Clausen, Conlon, Jager, and Metreveli 1978), is that disasters are class-based: what is a disaster for one social class (labor) need not be such for another (capital). The second, elaborated by Scanlon (1988), is that disasters create "winners" as well as "losers." Flood victims lose possessions, but relief agencies experience an increase in donations.

Exceptions can also become routine within a single social unit. They remain exceptions in the sense that unit members do not come to think of them as normal or everyday occurrences. However, the more frequently that exceptions of the same type reoccur and the more their features are alike from one occurrence to another, the more likely participants will develop routines to address them. In the case of illness, rituals and home remedies may develop. In the case of disasters, we have long called these routines for recurring exceptions disaster cultures (Moore 1964; Wenger and Weller 1973).

The same events may be exceptions in the short run but routine in the long run. This suggests the importance of understanding how social time is kept within routines. For example, a fifty-year flood is both exception (relative to the life-span of its victims) and routine (in terms of the statistical frequency of flooding in that flood plain). A Richter magnitude 7.1 earthquake is simultaneously exception and routine in geological time in terms of repetition cycles of sudden movements along the particular fault.

Exception routines are therefore routine in two senses: at the macro level, they coalesce into routine patterns of interactions and ideas for combating exceptions; and at the micro level, they become routine for the people who enact them. (Thinking of routines and exception routines in this way is related to Norbert Elias' concept of figuration; see Elias 1978 especially: 128–133.) In societies characterized by advanced division of labor, exception routines are enacted by formal organizations (fire departments, marriage counseling clinics, FEMA, etc.) which claim ownership of exceptions as part of their domains. Specialized occupational groupings and professions also claim, with varying degrees of success, the right to participate in if not to actually control exception routines. Whether the occurrence of exceptions attracts the attention of existing specialists or whether newly-emerging specialties create a recognition of new exceptions is an open question. (For empirical examples in which this question is raised, see, among many others, Nelson 1984: especially 11–19; Johnson and Hufbauer 1982; two systematic general statements are Cohen, March, and Olsen 1972; and March and Olsen 1979.)

The institutionalization of exception routines has several consequences. Because they are built around exceptions occurring in the past, exception routines are better suited to the most recent previous exception than to the present one. An example is military planning where strategy and tactics typically are based on the most recent previous war and often are inappropriate for the next war when it breaks out (e.g., French investment in the Maginot Line rather than an air force after World War I; American ground troops in South Vietnam in the 1960s, trained to confront a standing army but ineffectual in a guerrilla war). Thus the strength of institutionalized exception routines is also their main weakness. They may be splendidly effective against the "routine" exception (e.g., conventional fire fighting), but they may prove disastrous in even a slightly different setting (e.g., a fire involving toxic chemicals). New types of exceptions may go "undetected" because they lack the usual precursors monitored as part of exception routines. The "causes" of an exception may be "incorrectly" diagnosed because of the inevitable tendency to apply existing categories of thought and experience. An inappropriate exception routine may be invoked because its "owners" monopolize similar but unrelated exceptions, or no exception routine may be invoked because none is available.

Institutionalized exception routines change more slowly than do the exceptions which routines experience. Resistance to change comes both from powerful vested interests with a stake in the existing configuration of the exception routine as well as from the way knowledge and experience in a society is organized. Whether a new exception-routine structure will emerge to claim a new type of exception or an existing exception-routine institution will change to incorporate it is a question of organizational and professional power.

Even more interesting is the question of whether the emergence of exception routines is inevitable. Does every exception that disrupts social routines (structures) produce an institutionalized exception routine? A distinction between insiders (members of a social unit) and outsiders (non-members) is useful here (Merton 1972). It is possible that an outsider, with a different set of categories for organizing knowledge and experience, may identify as an exception something that unit members regard as routine and choose to do nothing about. Fatal illness is an example. If death is seen as part of the process of life rather than something separate and opposed to life, then death is more likely to be seen by insiders as routine rather than as exceptional. It is unlikely that rituals aimed at restoring health will develop within such a social system. The fact that a medical treatment for the illness may be available in some other social system is irrelevant. The key is whether system members themselves treat fatal illness as normal (routine) or exceptional (abnormal). This suggests the following two propositions.

When a distinction is made by members of a social system between what is routine and what is exceptional and when that routine is highly valued, an exception routine will eventually develop. The distinction between routine and exception is embedded in the world-view of system members, as is the value placed upon the routine and, as a consequence, the seriousness of the exception. Where the nuclear family unit of two parents and their children is highly valued, the rise of single-parent families will be treated as an exception and mechanisms proposed to restore the routine ("normal") family pattern. In other words, there will always be exception routines for addressing exceptions to routines that are highly valued.

With increasing structural differentiation, specialization means that exceptions tend to be covered by many different laws and to be allocated as responsibilities to many different organizations. Increasing differentiation and specialization also means a greater dependency of citizens on experts, professionals, and specialists of all kinds for the implementation of exception routines. Specialists organize to differentiate themselves from nonspecialists and to preserve their right to participate in or control exception routines. They may be threatened by nonspecialists' participation in exception-routine processes. Laypersons, on the other hand, may feel betrayed when organizations run by professionals are perceived to have failed to prevent or control disruption of routines.

Of special interest are exception routines sponsored by the state. Over time, the state—at least in so-called developed societies—has assumed more and more control over a variety of exception routines. Its monopolization of "security and public order" maintaining processes needs careful examination. A starting point is the state's claim to a monopoly of the right to use force (Weber 1978: 901–904; Weber 1958 [1946]: 78). The legitimacy of its claim is built into most social routines, but it is not without challenge. Most visible are groups such as the Freemen of Montana who defy many federal statutes and challenge the legitimacy of the federal government and other institutions. More widespread are everyday challenges to agents of the state such as suspicion of and hostility toward local police in minority neighborhoods, failure to report income on federal tax returns, withholding of information from government investigations and probes of various sorts, and thefts of government property (a denial that the state's claim to use material resources is more legitimate than taxpaying citizens' claims to use such property). (This list of challenges could easily be expanded by including legal challenges pursued in the courts.) These everyday challenges are different from direct confrontations such as revolutions. They are localized, isolated, and individual in character rather than collective, systematic, and public. They are usually labeled "lack of respect for the law." The point is that the state's claim to legitimacy does not go unchallenged.

Once a rule or law is adopted, to ignore its enforcement is to undermine the credibility of the state. For example, violations of the tax code are sought out and penalized not only for practical reasons (to acquire the revenue that the state feels it is legally entitled to), but also to shore up the erosion of public respect. In foreign affairs, a nation-state that declares itself to be a protector of freedom and a defender of peace loses respect as well as influence when it refuses to intervene in what other nation-states see as breaches of the peace and threats to freedom. This ongoing process of legitimatization (including de-legitimization and re-legitimization) is a starting point for analyzing the state's role in exception routines, including countermeasures following the disruption produced by disasters.

Briefly, the following propositions may be suggested. The state's principal role in dealing with disaster is to maintain order. One of the major functions of the state is to promote conditions under which the economy can succeed (among many others, see Lindblom 1977; Offe 1972). Political order and social stability are fundamental to the predictability required by businesses for decision-making and strategic planning. Government actions range from regulating economic transactions (to foster competition) to the creation and maintenance of the infrastructure (facilitating the production and exchange of goods and services). The state's disaster functions and its everyday functions are the same. That is, the state's disaster role is to minimize the disruption to economic routines caused by disaster (without adversely affecting business in the process) and to restore those routines as quickly as possible when they are disrupted.

Exception routines include such programs as small business loans, agricultural loans for crop damage, funds for the repair of public infrastructure such as streets and highways (people need to get to work, shoppers to stores, raw materials to factories, goods to wholesale and retail locations, etc.), funds for repair of educational buildings, and funds to "get families back on their feet" (i.e., earning wages and paying taxes once again). The nature of these exception routines is an observable indicator of the importance of the routines that they are invoked to restore.

In general, routines, exceptions, and exception routines are all interrelated parts of the same process. Each part is as natural and as normal as the other parts; none is at all unnatural or abnormal in the sociological sense. Therefore, I propose to refer to all three as comprising the process of routinization. Logically, a routine must first coalesce into a set of structured relationships. Put differently, for something to be a disruption or breakdown there must first exist something to be disrupted or to breakdown. Furthermore, a single exception (disruption, interruption, or breakdown) may be dismissed as unique, a fluke. However, routines which persist will exhibit repeated exceptions. Exceptions themselves become "routine." The common elements across several exceptions become the model for developing exception routines. As elements of the same social process (i.e., the process of routinization of social structures), neither routines, exceptions, nor exception routines are static. In fact, changes in routines produce changes in the nature of exceptions. Since change in exception routines takes place more slowly, there is always a lag between exceptions and exception routines (i.e., the cliché "armies always fight the previous war"). Each new exception is potentially more likely to be at variance with an exception routine.

The introduction of these concepts of routines, exceptions, and exception routines is not intended to constitute another attempt at a middle-range theory of disaster. Rather, it is intended to be a link between the empirical study of disasters as well as other types of exceptions and a general theory of social order. Many sociologists provide a foundation for the continued development of such a theory using empirical findings from the study of disasters and other exceptions. I will mention only a few.

Durkheim's work on social solidarity (1964), although skewed toward the connection between the individual and a preexisting social order, is an obvious starting point. Weber's writings on the modern state and legitimate authority (1978: 901–1005) are also fundamental beginning points on the process of legitimization. More recently, Giddens' work on late modernity (1990, 1991) places questions of social solidarity and legitimization in a contemporary context. Luhmann (1993) provides a vocabulary for analyzing how highly differentiated social systems attempt to deal with the uncertainty created by disruptions. His own analyses examine the language of risk and danger as part of a general social process by which social systems reduce complexity in their environments. His general work on systems theory (1992, 1995) remains an untapped resource for disaster researchers.

Lars Clausen (1978, 1983, 1988, 1992) presents a systematic attempt to integrate disasters into a general theory of social differentiation. He argues that disasters:

> are the normal outcome of social change. Certainly unwelcome they are (other than, perhaps, revolutions), certainly rare (but in no way rarer than revolutions), but even more certainly, as normal as the first prize in a national lottery. Normal social events must be provided for by general sociological theory.
>
> (1992: 182)

Clausen subsumes disasters within a general theory elaborated through the FAKKEL process-model of disasters (1983; for an English version, see Clausen 1992). FAKKEL represents the first letters of the German words for the six stages of this model: *Friedensstiftung* (initial foundation of peace resting on the mutual solution of basic social problems); *Alltagsbildung* (the establishment of everyday routines); *Klassenformation* (formation of classes and rising class struggle); *Katastropheneintritt* (onset of catastrophe); *Endekollektiver Abwehrstrategien* (the unconditional surrender of collective defense); and *Liquidation der Werte* (collapse of the common value structure). Clausen places special emphasis on the increasing differentiation and conflict between problem-solving experts and laypersons as a source of growing vulnerability to disruption (especially Clausen 1988).

All these writers provide exemplars of how fundamental sociological questions can be explored equally well with disasters as the objects of attention or with other topics, as circumstances dictate. For the discipline as a whole, questions about disasters are less important than questions about societies that can be answered by studying disasters.

CONCLUSION

Limitations of space do not permit me to say more than a sentence or two about another implication for research emerging in the earlier IJMED papers. An entirely different research program is implied by the recognition of the importance of existential definitions of disasters. This program would involve studies of how the term disaster is used to organize knowledge, experience, and action (including choosing between action and inaction, or noninvolvement). It would entail empirical studies of legal definitions of disaster; how the press and other news organizations define disasters and how such definitions affect the deployment of organizational resources, coverage, etc.; formal and working definitions of disaster as they affect the collection and organization of information (e.g., how disaster is defined and used in library systems, encyclopedias, and computerized databases; definitions of disaster in popular use; definitions in various academic disciplines; and what constitutes a disaster in the arts including film (Quarantelli 1980), fiction, poetry, nonfiction, painting, and music. These are not proposed as "fun little studies" that one could do on the side. If one is serious about answering the question "What is a disaster?", then the way to proceed is to find out how the people we study would answer this question.

On the main point of this chapter, the call for a return to fundamental sociological questions answerable by the empirical study of disasters has several potential drawbacks. One is that research funding sources are not interested in questions of theory (Dynes and Drabek 1994; Quarantelli 1994, 1995b). They have important practical questions related to their missions. They also face threats of downsizing, budget reductions, and sometimes proposals for their virtual elimination. It is therefore unrealistic to expect that research funders will look favorably upon proposals that are easily perceived as "too theoretical" or "too abstract."

Some options do exist. One is for us to work even longer hours, doing the kind of studies that clients of our research want but afterwards trying to make explicit the connections between our findings and core theoretical questions. Another option, one that has not been systematically explored, is to engage much more frequently in what university administrators sometimes call "unfunded research." That is, we should explore gathering and analyzing materials at our own expense. Trips to conferences, for example, can become opportunities to visit nearby archives, libraries, and newspaper offices. Stops can be arranged along the way to conduct interviews and gather materials. Vacations can be built around visits to important research sites (subject, of course, to objections from spouses and family members!) The larger point is that we should think about ways that we can carry out our intellectual pursuits with less outside support, or even no support at all.

Another potential downside to the recommendation in this chapter is that the timing may be bad. With cutbacks in both university faculty and curricula, calls for increasing attention to theoretical concerns and for exploring what can be accomplished through unfunded research may be detrimental to the short-term interests of the discipline. However, in the long run it is the body of knowledge that sociologists possess that will determine the future of the discipline within the academy. If all we have to "sell" are the methods by which we are known (e.g., survey research), then the future of the discipline will be determined solely by the availability of extramural research support.

Finally, it is worth asking a fundamental question: "Why study disasters at all?" There is only one defensible answer: The empirical study of disasters is a means for answering important questions. The suggestions in this chapter have not been calls for radical changes in what we do. Instead, they are recommendations about how we do what we do. We should return to a focus on the central questions of sociological theory. These are questions about how societies distinguish the typical from the atypical and about how they attempt to steer the latter back toward the former. Stated more abstractly, the theoretical questions, as they have always been, are about how societies try to maintain order in the face of uncertainty, about what they treat as essential to order, and about the consequences of restoring order. The future of the discipline will be determined by how successful we are in answering such questions.

NOTE

1 This paper was written while the author was a visitor at the Katastrophenforschungsstelle, Christian-Albrechts-Universitate zu Kiel. My thanks to Wolf Dombrowsky, Lutz Ohlendieck, Henrich Fenner, Jutta Kast, and Lars Clausen of the Institut für Soziologie at Uni-Kiel for various forms of support, both material and intellectual.

13

FUTURE DISASTERS, FUTURE DEFINITIONS

Uriel Rosenthal

INTRODUCTION

One of the most prominent scholars in disaster sociology recently admitted that for more than four decades he has "struggled with how to define and conceptualize the term 'disaster' " (Quarantelli 1995e: 221). At first glance, this might read almost like a confession. But on reflection, this candid observation reveals a firm belief in academic openness. One should indeed cherish the distinct willingness to accept different perspectives and constructs of what is a core concept in this field of research. Coming from a scholar who has an intimate knowledge of the empirical domain of disasters, this appeal for conceptual clarity is meant to strengthen the underpinning of empirical research, and definitely not "to engage in some useless or pointless academic exercise" (Quarantelli, 1995e: 225).

The continental-European tendency to look for, and indeed construct, the "essential attributes" of social phenomena may easily lapse into the fruitless academism referred to above (Gregor 1971: 133). The pursuit of logical coherence and comprehensive conceptual frameworks runs the risk of falling a prey to such academism. It may end up with beautiful schemata and may solicit well-deserved admiration for acute thinking, but the static logic of essentialism may well be out of line with the dynamics of social reality.

Quarantelli proposes that we look at the *characteristics* as well as the *conditions* and *consequences* of disasters:

> to be concerned about what is meant by the term "disaster" is not to engage in some useless or pointless academic exercise. It is instead to focus in a fundamental way on what should be considered important and significant in what we find to be the characteristics of the

> phenomena, the conditions that lead to them, and the consequences that result.
>
> (1995e: 225)

This compels us to pay attention to the dynamics of disasters. Consequently, at the very beginning, the terms of reference for our exercise take us well beyond the traditional notion of disaster as an event with a strictly demarcated start and finish. We should benefit from this process-oriented perspective and should, for that matter, try to match the concept of disaster with empirical developments and trends in the world of disasters and crises.

In order to underscore the importance of a concept of disaster which corresponds with empirical developments and trends, we structure our observations into three parts reflecting on the past, the present and the future: the traditional world of disasters, contemporary views, and disasters and crises to come. The main focus is on the interplay between empirical developments and their conceptualization. In doing so, our approach comes close to an attempt to apply the basic notions of the sociology of knowledge to the world of disasters and crises. We expect conceptual developments to be intimately linked to social dynamics (Merton 1968: 510–562).

CHARACTERISTICS, CONDITIONS, CONSEQUENCES: THE TRADITIONAL WORLD OF DISASTERS

The traditional world of disasters is one of natural agents, sudden onset, and a well-defined and clearly demarcated context. It is the world of "un-ness." The *characteristics* of disasters are unequivocally negative.

> Disasters are unmanaged phenomena. They are the unexpected, the unprecedented. They derive from natural processes of events that are highly uncertain. Unawareness and unreadiness are said to typify the condition of their human victims. Even the common use of the word [disaster] "event" can reinforce the idea of a discrete unit in time and space. In the official euphemism for disasters in North America, they are "unscheduled events".
>
> (Hewitt 1983a: 10, 1995: 332)

Since the beginning of this century, the scope has been somewhat widened, as it now encompasses man-made disasters such as explosions and transportation accidents. But, in a way, explosions and transportation accidents only emphasize the characteristic features of traditional disasters (Prince 1920). The "un-ness" of such man-made disasters manifests itself in its conspicuous challenge to the positivism of modern technology. Technological designs that are seen as the evidence of human control over natural forces, turn out to be highly vulnerable to calamitous interventions.

In the traditional view of disasters, natural agents and technical failures hit the immediate site and the vicinity of the accident with tremendous force and without delay. Initial endeavors in disaster research have paid ample attention to the damage inflicted upon the affected parts of the social system. Nevertheless, in the traditional approach, practitioners as well as analysts are inclined to focus on decisions and measures which help to reduce the timespan and the social and political consequences of the unfortunate event. The main preoccupation of traditional disaster management and conventional disaster research is with a prompt return to normalcy. Society should get back to normal. Disasters may produce change, but the dominant perspective relates to how the social system is able to overcome the strain from exogenous abnormal forces. In the typical vocabulary of systems analysis, disasters may generate change *within* the social system; they are, however, not supposed to cause systemic change (Parsons 1951).

All in all, Fritz's classical definition of disaster more or less exemplifies the traditional view of disasters:

> an event, concentrated in time and space, in which a society, or a relatively self-sufficient subdivision of a society, undergoes severe danger and incurs such losses to its members and physical appurtenances that the social structure is disrupted and the fulfillment of all or some of the essential functions of the society is prevented.
>
> (1961: 655)

It could be argued that this definition does not give full credit to the original notion in the social sciences of disaster. Some of the founders of modern sociology indeed combined ideas of discontinuity, crisis, and disaster with explorations of social change (Sorokin 1928). With regard to the origins of disaster research, Dombrowsky (1995a: 247–248) rightly points to Carr's observation that disasters can only be understood as the cultural dimension of a catastrophic event (1932: 207). However, here too the dominant perspective is system-oriented. The system is blamed for its apparent incapacity to restore social stability and order. To the extent that this disaster perspective already allows for a dynamic idea of disaster, the course of events is bound to take place well within the boundaries and constraints of the existing social system.

In the traditional view of disasters, two categories of *conditions* appear to be dominant. Self-evidently, the scourge of God together with social or political negligence have traditionally served as the principle conditions of natural disasters. Gradually, negligence has given way to more specific conditions such as deficiencies in mitigatory policies and preparatory measures.

With regard to man-made disasters, the original explanation usually emphasizes either technical failure or human error. Here the very problem is the occurrence of disasters in those domains of human intervention which, right from the beginning, have created the illusion of control and mastery over the forces of nature. In a way, the problem lies with the antithesis of modernity as the control of man over hitherto uncontrollable forces, and disaster and crisis as the cogent evidence of man's fallibility.

It is no wonder, then, that initial explanations, if not justifications, of such ostensibly "illogical" disasters as catastrophes in airways and aeronautics rest with the bad luck of flaws in technical design and with the unpredictability of human nature. To avoid any misunderstandings as to what is the contemporary relevance of this traditional view of disasters, it should be added that a great many official investigations as well as public opinion still cling to technical failure or human error as the number one cause of man-made disaster. In determining the conditions of disaster, technical failures often take its place as an appropriate substitute for the act of God, whereas human error reflects the inherent weaknesses of mankind and, for that reason, solicits empathy with the unfortunate materialization of abstract risks into concrete, physical calamity (Horlick-Jones 1996: 62–63). If one takes seriously the relevance of a sociology of knowledge for the conceptualization of disaster, one might say that there must be other explanations—both politically and organizationally—for the pertinency of traditional considerations in these (ultra) modern sectors of technology.

Looking at the *consequences* of disasters, the traditional view of disasters is quite distinct. The classical response to natural disasters, which linked with a fatalistic idea of conditions and causes, was simply to give up the stricken area. In its mild version, the negligence on the part of the social and political elites regarding the weaker parts of society culminated in retarded and opportunistic measures to cope with the damage inflicted on victims of disasters.

It should be stressed that here too contemporary disaster research can benefit from this seemingly traditional approach. Up to now, differences in disaster response, and even more so in disaster recovery and rehabilitation programs, may still be explained from the political relevance and leverage that elites attach to areas that have been hit by the disaster. Examples run from the differentiated response by the Italian political elite to earthquakes in northern and southern Italy (Geipel 1982), to the political considerations in Washington DC to issue a federal disaster declaration (May and Williams 1986: 6–7).

The limits of the traditional concept of disaster

The traditional concept of disaster stresses a narrow time–space setting. Agents and agent-driven events are definitely more powerful ingredients than the chain of events existing for conditions, characteristics and consequences. In the traditional view, the conditions of disaster are confined to either metaphysical causes or simple categories of technical and human error. The empirical scope remains limited to a fairly obsolete perspective of purely natural disasters, and to a restricted range of proto-technological hazards. The concern with the consequences of disaster usually does not reach beyond short-term considerations. If, by exception, long-run consequences are taken seriously, the argument is rather mechanistic. According to the traditional view, disasters only bring the structural and cultural shortcomings of the social system into the open, thus contributing to its downfall. Clearly, the traditional sociology of disasters fits in with the traditions of structural functionalism in sociology at large.

CHARACTERISTICS, CONDITIONS, CONSEQUENCES: THE CONTEMPORARY WORLD OF DISASTERS

The contemporary world of disasters is a complex world of linkages, chains, and processes. For that matter, the *characteristics* of contemporary disasters are best phrased in terms of the corresponding adjectives: complex, linked, connected, processual. What looked as an omen for the future in Dynes's classic of 1974, is now an integral part of the world of disasters: his type V consisting of "compound disasters" (Dynes 1974: 63). Taking into account Quarantelli's plea for the study of characteristics as well as conditions and consequences, one can argue that by now it is just necessary for disaster studies to link conditions, characteristics, and consequences. Any other conceptualization and the empirical approach will simply fail.

Let us pursue this argument. The classical distinction between natural and man-made disasters has given way to more advanced thinking. Geographers and geologists, as well as sociologists and political scientists, agree upon the fact that floods, droughts, and other so-called natural disasters relate directly to human intervention into the ecological system as well as to outright technological interference and human exploitation. Instead of the necessity of nature, human factors and social forces involving choice, decision making, and management come to the fore. As Hewitt says:

> social understanding, and socially just and appropriate action, are the more crucial issues for the contemporary disaster scene.
>
> (1995: 318)

Furthermore, the concept of compound disasters sharpens our concern with disaster agents which start as seemingly clear-cut instances of a distinct category of calamitous events, but subsequently, as it were, unintentionally evolve into a totally different category of disaster or crisis. In quite a number of cases, the genuine predicament of disasters is not so much with the number of victims and the *objective* amount of material and immaterial damage as it is with the social and political turmoil and the unleashing of hitherto latent forces. In a complex dynamic process full of intricate interactions and often perverse effects of well-intended decisions and measures, the disaster agent loses ground to disaster context.

> *The Amsterdam air crash of October 1992.* On Sunday, October 4, 1992, a Boeing 747 freighter, out of control, bored its way into two apartment blocks in Amsterdam. The crash cost 43 lives, including the aircraft's crew of four. Emergency services came in vain to the site of the accident. There was no way whatsoever to try and rescue people from the inferno and there was only a handful of wounded people. With the benefit of hindsight, it can be argued that the disaster, which at first sight was one among so many aircraft accidents, would better be regarded as an urban crisis. The authorities' willingness to provide ample emergency relief, in time gave rise to totally unexpected tensions within and outside the multi-cultural district where the Boeing had crashed. It actually turned the disaster into an acute issue concerning illegal immigration and abuse by pseudo-victims—both illegals and legal inhabitants—of the authorities' creed to caring government. So, rich man's aircraft disaster ended as a poor man's crisis.
>
> (Rosenthal, 't Hart, Van Duin, Boin, Kroon, Otten and Overdijk 1994)

Looking at the *conditions* of contemporary disasters, we should move away from metaphysical origins and simple modes of technical failure and human error. In many ways, contemporary disasters reflect the complexity of modern society and present-day technology. Many contemporary disasters follow the lead of complex technology, common-mode design, and multi-functional use of infrastructures. According to some observers, certain kinds of high-risk technologies indeed create conditions for "normal accidents" (Perrow 1984). Another powerful notion concerning the conditions of contemporary disasters is Turner's conception of incubation, that is:

> the accumulation of an unnoted set of events which are at odds with the accepted beliefs about hazards and with the norms for their avoidance.
>
> (Turner 1978; see also Toft and Reynolds 1994 and Turner and Pidgeon 1997)

Seen from this angle, simple technical flaws and human errors inevitably lose prominence as primary conditions of disaster. The focus shifts to the organizational and systemic level. Operational errors should be seen in the broader context of a deficient safety culture and lack of effective counter-disaster management.

The main ideas about the conditions of contemporary disasters take us well beyond act of God, sudden onset, and sheer surprise. To be taken by surprise cannot serve any more as a sufficient excuse for the prolonged and protracted negligence of long-term adverse trends and equally long-term passivity. The concept of creeping disasters catches such long-term conditions of disasters. It relates to a realm of phenomena that seem to be totally opposite to the traditional world of disaster agents and disaster events. Deforestation and desertification, soil salination and fertilizer use, the losses of ozone, global warming and the rise of the sea-level span many decades (Karplus 1992). It is abundantly clear that the classical concept of disaster does not cover these categories of disaster. Creeping disasters may of course burst into the open at a specific site, on a specific moment. But with such long-term developments it would be wrong to focus on the classical dimensions of territorial specificity and sudden onset.

In comparison with the traditional view of disasters, the dominant view about contemporary disasters is that more often than not, they have pervasive *consequences*. The so-called "disaster after the disaster" lends a new dimension to the traditional idea that the post-disaster era is best described in terms of recovery and rehabilitation. The "disaster after the disaster" may have a stronger impact than the apparent disaster. Communities that get struck by a natural or man-made disaster may lose their sense of community to the extent that a recovery of communal identity is highly unlikely (Erikson 1976, 1994).

This way of looking at disasters takes us an additional step away from the traditional view of disaster as a concrete event which comes by surprise and which only necessitates a one-way process back to normalcy. There is no reason why the aftermath of a disaster could not take on critical proportions worse than the disaster itself. One should be careful not to get enmeshed in a game of words that would reserve the label of disaster for the initial event, and which would call subsequent problems and intricacies by the socially or politically correct phase of "recovery" or "rehabilitation."

The complexities of the contemporary concept of disaster

The contemporary concept of disaster actually satisfies the need for a dynamic perspective relating characteristics to the conditions and consequences of disaster. The interrelationship between characteristics, conditions, and consequences is the best guarantee for a process-oriented approach. That approach should be critical for the compelling sequence of official models which, for understandable reasons of managerial clarity, ordain disaster response to be preceded by mitigatory and preparatory policies, and, in a more compulsory manner, to be followed by disaster recovery and rehabilitation (Petak 1985). Contemporary notions of disaster should encompass various patterns of disaster development. For instance, what at the onset seems to be a genuine disaster may turn out to be only a minor accident in comparison to catastrophic events and developments yet to come. From a political and managerial perspective, it should also be understood that good intentions in disaster and crisis management sometimes make for bad consequences. Deficient emergency responses may turn an accident into a fully fledged disaster.

The Sandoz chemical accident of November 1986

The Sandoz case started as a fire in one of the plants of this pharmaceutical corporation in Basel: the accident would never have reached the proportions of a large-scale disaster were it not for the totally counter-effective response on the part of ill-equipped and badly trained firesquads.

The Eindhoven aircrash disaster of July 1996

In July 1996 a military Hercules air carrier transporting more than thirty members of a military brass band crashed on the military airport of Eindhoven in the Netherlands. Most people on board survived the impact of the crash. But due to misunderstandings between air traffic control personnel and emergency personnel, the emergency services assumed there had been only a few crew members on board who must have died in the fire which broke out immediately after the crash. Efforts to rescue the people on board started only well over a half hour after the crash, when emergency workers encountered a shockingly large number of bodies within the aircraft (Crisis Research Center 1996).

Decisions regarding the allocation of emergency equipment and personnel may help to manage one disaster, however, at the expense of an even worse disaster yet to come. Contemporary disasters are no exceptions to the law of requisite variety. Complex disasters require complex decisions. Mechanical models of action and thinking will not meet the complexities of the linkages, chains, and open-ended processes of contemporary disasters.

URIEL ROSENTHAL

CHARACTERISTICS, CONDITIONS, CONSEQUENCES: FUTURE DISASTERS AND CRISES

The social sciences tend to be rather conservative in their conceptual and terminological endeavors. Definitions and concepts usually recur to empirical referents of past and present phenomena and trends. In their appreciation of the theoretical significance of a concept, social scientists focus primarily on its usefulness in describing and explaining the real world as it manifests itself in the past and the present. Only seldom do definitions and concepts encompass future developments. Some watchers of the future indeed prefer to use a distinct terminology and construct special concepts to emphasize the difference between contemporary and future trends.

The relevance of our explorations into the definition and concept of disaster depends, however, to a considerable degree on their contribution to our understanding of future disasters and crises. Attempts to define and conceptualize the notions of disaster and crisis should not only look at the past and the present, but should also take into account the possible characteristics, conditions, and consequences of future disasters and crises. It is a fascinating challenge to try and match the concept of disaster with future trends and developments in the world of disasters and crises. If we were able to formulate a concept of disaster which would stress the basic dimensions of future disasters and crises, together with a typology derived from a future-oriented disaster concept, this would help us a lot to build a research agenda for the years to come.

Transnationalization

Disasters and crises to come will feature transnational and even global characteristics. The original source of the problems at hand may continue to be local or national, but the immediate and long-term impact of disasters and crises to come will span countries and continents. A great many major disasters and crises of the last decade already indicate the significance of this transnational dimension. The Chernobyl disaster of 1986 foreshadows the kinds of disasters which high-risk technologies may produce. Until the Chernobyl disaster, nuclear power plant disaster scenarios were typically nation-bound, with international implications being subordinated to national concerns. But the radiation fallout from Chernobyl caused substantive damage to many countries in northern and western Europe. Interestingly enough, those countries may throughout the postwar period have experienced the proximity of the former Soviet Union in the domain of military security, but they felt themselves at safe distance from Soviet nuclear energy plants. The ironical fact was that, apart from the two Berlin crises (in 1948 and 1961), Scandinavia and Western Europe did not suffer a genuine threat in the traditional domain of international crises, but eventually were exposed to immediate danger due to an unexpected accident in a nuclear power plant in the Ukraine. Since the dismantling of the Soviet Union, one of the most serious worries of the Western countries—both the United States and Europe—has been the very bad state of nuclear power facilities in Russia and the other Eastern European states.

Environmental disasters are another category of disasters that call attention to transnational impact. The most cogent set are the kinds of ecological trends which indeed span and, according to some experts, threaten the Planet Earth. A summary of immanent environmental catastrophes deals with depletion of the ozone layer, global warming, low-level radiation, acid rain, each asking for international actions. The estimated costs of proposed measures amount to astronomical numbers: "billions and billions of dollars," "5% of national defense budgets," "trillions of dollars in the United States alone," "many billions of dollars in the United States, Europe, and Japan" (Karplus 1992: 268–269).

The dominant paradigm of long-term creeping disasters reaches beyond the still contained notion of external effects which describe the linkage between localized origin of a threat and damage inflicted upon another specific location. Nevertheless, the future will see an increasing number of disasters and crises which highlight in a very concrete and tangible manner the obsoleteness of national borders. We will see more and more disasters like that of the Sandoz chemical accident in Basel, Switzerland (November 1986), which, partly due to deficient emergency measures, produced considerable pollution along the Rhine River over more than 800 miles. It speaks for itself that a river that runs through several countries, does not declare polluted water at the customs.

The missing barrels of Seveso, 1983

Another foreboding of what the world of disasters and crises will come to look like is the post-Seveso crisis of the missing barrels (Lagadec 1988: 135–139). Following the accident at the Hoffman-Laroche chemical plant in the Italian village of Seveso in 1976, the so-called Seveso Directive, which was issued formally in 1982, became the symbol of the European Union's effort to develop Union-wide policies on industrial safety. But the ink of the Seveso Directive had not yet dried up when the European countries had to face the frightening case of the Seveso barrels. Between March and May 1983 the European countries had to engage in a frantic search for 41 missing barrels containing 2,200 kilos of highly contaminated materials of the Seveso plant. Instead of turning into the symbol of European industrial safety policy, Seveso suddenly emerged again as the prototype of a crisis which despised national borders and reacted a nerve-racking situation throughout western and southern Europe. After two months of commotion, the barrels were found in the backyard of a slaughterhouse in northern France. Meanwhile, the crisis had imposed itself on the authorities and public opinion in nearly all European countries, from Italy and France to Western and Eastern Germany.

The transnational dimension extends to other categories of future disasters and crises. The adverse consequences of hitherto nation-bound conditions such as ecological deterioration, economic stagnation, as well as urbanization (mega-cities) and hyper-industrialization, increasingly affect the social and political situation in adjacent countries. In Africa, mass migration and refugee movements reflect the artificial and conflict-ridden nature of state boundaries. International aid and relief operations may seem to be the logical answer to this kind of transnational misery, but, for that very matter, will easily become subject to accusations of favoritism and politicking. It has indeed been shown that, if delivered without sensitivity to such transnational intricacies, international assistance may aggravate rather than reduce the plight of the people concerned (Benini 1993: 215–228).

Mediazation

The subjective notion of disaster and crisis can be summarized in a version of the Thomas theorem: "If men define a situation as a crisis, it will be a crisis in its consequences" (Thomas and Thomas 1928: 572).

Undoubtedly, social science must feel comfortable about this subjective definition of disaster and crisis (Porfiriev 1995: 288). Damage and disruption do only count to the extent that they are indeed regarded as a threat to society and its members.

When citizens or authorities define or declare a difficult situation a disaster or a crisis, this may have a decisive impact on the subsequent course of events. To call the situation by such an evocative word may provoke hyper-vigilance and overreaction, running from collective stress to escalatory decisions on the part of the authorities. On the other hand, it may solicit the collective energy and mobilize the emergent rules and norms necessary for effective disaster and crisis management.

For the years to come, the subjective notion of disaster and crisis will become even more important. For future disasters and crises will increasingly follow the media-instigated lead of the Thomas theorem: If the media define a situation as a disaster or a crisis, be sure that it will indeed be a disaster or a crisis in all its consequences. In conjunction with transnationalization, future disasters and crises can best be caught in the CNN red-colored headline: Site in Crisis—be it the Gulf area, Israel, Rwanda, Oklahoma City or San Francisco. Mediazation will be one of the driving forces in the world of future disasters and crises. The media can bridge the distance between the original crisis site and all those other sites in the world that, though hundreds or thousands of miles away, are an integral part of the context at hand. This holds for the crisis centers in the main capitals of the world, as well as for millions of people that experience intense anxieties and sentiments in the slipstream of world news. The transnationalization of the media makes them a most formidable player for the very reason that, in apparent contradistinction with national powers and international agencies, they are able to cover every and all corners of the ever-expanding domain of disasters and crises (McQuail 1993: 292). They seem to have easy access to the various sites that together make up the crisis context.

The media, television in particular, suggest a bridging of the distance when, in fact, they may very well widen the gap between those at the site of the disaster and those at home or in the crisis center. The Gulf War foretells how the media may color or discolor the facts and figures of the crisis. According to Taylor:

> the Gulf War has presented a new challenge: the public's apparent desire *not* to know beyond the sketchiest details what is going on while it is going on. Whereas journalists see speed as essential to their profession, their readers seemed more than willing to wait until the military could report that a mission had been accomplished before finding out about it.
>
> (1992: 274)

Basically, then, mediazation hammers home the subjective mode of disasters and crises. It may go all the way from outright sensationalism to self-imposed censorship. To the extent that the media play an important role in defining the situation, on both ends of the spectrum this underscores the distinction between what is really going on and what is brought to bear upon the decision-makers, emergency workers, the people in stricken areas and the public. So mediazation does not accommodate too well the traditional definition of disaster and crisis in terms of objective measures of harm (number of people killed and amount of material damage). It fits better into a new category of disasters and crises which is characterized by extreme collective stress rather than fatal casualties or significant physical damage. In such cases as the medical radiation contamination in Goiana, Brazil (1986), as well as various casualty-free kidnappings and hijackings, "We need to get away from equating disastrous occasions only with fatalities" (Quarantelli 1996: 232).

At the same time, however, mediazation may help to bridge the gap between the "objective" definition of past disasters and the highly "subjective" one of disasters yet to come. For the time to come, the media will sort out for intensive coverage two categories of disasters. On the one hand, they will be keen on the ominous prospect and occurrence of mega-disasters. This category of disasters defines itself by the sheer number of potential or actual casualties and the astronomical amount of physical damage. Apart from the death toll of civil strife and the ordeal of mass refugee movements, this applies to mega-carriers in the transportation industry—from the foreboding of the Estonia ferry disaster with nearly 900 victims (in September 1994) to the grim prospect of the new 600–800-passenger Boeing planes—as well as to nuclear power plant disasters which, though occurring only once in 33,000 years, will inflict an astronomical damage upon the affected countries (Lauristin 1996: 85–94). On the other hand, the media take a particular interest in the typically subjectivist category of disasters and crises that, without too much reference to tangible referents, feature craze, panic, and collective stress. Such disasters and crises may put a heavy burden on the social fabric. They may weaken the normative structure of society.

Politization

The just-mentioned trends prepare the ground for a marked politization of future disasters and crises. The transnational dimensions of disaster and crisis challenge the credibility and legitimacy of the politicians. Transnationalization may easily be regarded as final proof that the politicians are no longer in control of what is going on in the world of governance and management. The obvious argument that the cross-border origins of transnational disaster exonerate them from domestic responsibility does not hold. After all, part of the burden and damage will be inflicted upon their own people and, even more importantly, sooner or later their own country may be the source of an expanding disaster and fall a prey to the scorn of other countries. More than ever, "it cannot happen here" is a high-risk aphorism.

The prominent role of the media also contributes to the politization of disasters and crises. The media speed up the political process of disaster and crisis management. They put pressure on the decision-makers and crisis managers to explain and justify what they do in order to avert the threat or bring the crisis to an end. They make it increasingly difficult for the authorities to suspend their public appearance until the crisis is over. On the contrary, the media will compete to be the first to confront the political authorities with incisive questions about possible causes, deficient emergency operations, and litigation procedures. If the political authorities are hesitant to inform the public, the media will find others to give the answers: such as emergency workers, whistle blowers, and experts. This will not strengthen their political position.

Disasters and crises will easily turn into high politics. The context of future disasters and crises will best be depicted as a context of framing and blaming ('t Hart 1993: 36–50) It requires quite some political skill to turn disaster and crisis from threat and adversity into an opportunity and a political asset.

> *The Challenger disaster (January 1986)* Ronald Reagan showed all the necessary skills in the immediate aftermath of the Challenger

> disaster. The U.S. National Aeronautics and Space Administration being the "bad guys," it was only after several hours of political silence that the American President went public to express the feeling of grief and sorrow in the name of the American people. In that way, he uncoupled himself from the independent agency of NASA. A recent analysis of the Challenger disaster ironically suggests that more than high-powered political and organizational pressure, the cause of the tragedy was "a mistake embedded in the banality of organizational life."
>
> (Vaughan 1996: xiv)

CONDITIONS AND CONSEQUENCES: THE CIRCULAR MODE OF FUTURE DISASTERS AND CRISES

The three dimensions of future disasters and crises—transnationalization, mediazation, politization—squeeze conditions and consequences into a circular mode. The transnational aspects may allude to a separation between national conditions and international consequences, but it would be naive to embrace that line of thought. It does not take too much effort to put part of the blame for one country's predicament on its neighbors or on the international community at large.

Being confronted with the calamitous impact of a disaster across borders, the authorities may well be pressed by the media to elaborate on their own lack of preparedness and to state publicly what everybody wants to hear: "This cannot happen here". Then, however, there will immediately be others to renounce that claim.

In the political arena, responsibilities will not be confined to the domestic scene. In many parts of the world, the national authorities govern what has been called a semi-sovereign state. Increasingly, political responsibilities are shared under the aegis of international treaties and transnational agreements. Across borders, the origins of disasters and crises will merge with their consequences.

Conceptualizing future disasters and crises

The study of future disasters and crises will benefit from a concept that stresses the *transnational course of events*, *underlines the subjective dimension of mediazation*, and *brings politics back in*. Future disasters and crises will span the entire spectrum running from mega-disasters, that will cost the lives of many people and inflict substantive physical damage, to instances of extreme collective stress that emanate from widespread rumors, hypes, and incidental atrocities.

14

LEGISLATORS, INTERPRETERS, AND DISASTERS

The importance of how as well as what is a disaster

Steve Kroll-Smith and Valerie J. Gunter

We were invited to comment on the six papers that appeared in the special issue of the *International Journal of Mass Emergencies and Disasters* (IJMED) each addressing the question "What is a Disaster?" We were also asked to offer our own answer to this question. It will become clear to you as we proceed that we do not think there is one answer to this question, or indeed, just one question to ask. We do, however, think we are engaged in a worthy discussion in so far as it is an occasion to talk frankly about sociology and disaster.

We begin with two observations we made while reading the special issue. First, while we have more data about disasters than at any other time, we are also less certain about just what a disaster is. Indeed, there is almost an inverse (and seemingly perverse) logic at play here: the more we know about specific disasters the more definitions of disaster are registered in the literature. And second, it appears sociologists who work within the classic model of disaster research are not seriously considering work by other social scientists who are not a part of the classical tradition, and thus risk being overtaken by developments that according to their paradigm ought not to exist.

Both of these observations are, in our opinion, derivatives of a broader more inclusive observation. To wit, the current debate over what is a disaster can no longer be separated from the broader, contemporary debate over what is sociology. To continue to do so is to risk conversing amongst ourselves in the absence of the good insights of others who are attempting to clarify just what it is that sociologists do. By paying attention to this more inclusive debate we should not be surprised to find that the specific debate over the boundaries of a sociology of disaster is, in fact, a subtext of a noisy assembly of voices arguing their version of the boundaries of the discipline.

A good place to begin is with the question "What is sociology?" and its relevance to our more immediate question, "What is disaster?" From this discussion two stances toward sociology and disaster are identified: the legislative and interpretive. The definitions of disaster in the special issue are shown to be legislative. A case is made for the importance of the interpretive stance in complementing the question "What is disaster?" with the more awkward question, "How is disaster . . . ?" Examples from our research and others are used throughout the discussion. A conclusion reveals what some might consider a sinister plot to plant a few weeds, a little "wild knowledge," in the otherwise exquisite garden of classical disaster sociology.

WHAT IS SOCIOLOGY? WHAT IS DISASTER?

Reading the six good papers on the definition of disaster and considering the historical evolution of the topic, we are struck by how this debate is occurring without acknowledging the more inclusive controversy in sociology regarding the very parameters of the discipline. In our opinion, this disciplinary dispute can clarify key problems we are having with our more parochial question, "What is disaster?" In particular, we think it will show that we are stumbling over the more essential question of how to define the problem of definition. The strategy to define the problem is itself, in our opinion, part of the problem. Consider first the broader question, what is sociology?

As we approach the end of a century founded on what some call the "charisma of reason" (Whimster and Lash 1987; Smart 1993), several respected sociologists are questioning the validity of a unified discipline founded on standardized definitions, experimental methods, and an incrementally coherent body of knowledge (Bauman 1987; Giddens 1987; Denzin 1992; Touraine 1995). Other respected sociologists, however, are defending sociology's original agenda arguing that theories of behavior applicable to all times and places are possible and desirable, indeed is the reason we gather together under the banner, sociology (Haynor 1990; Wallace 1991).

Is sociology a discipline in which theories, concepts, and methods are precision instruments and "measure(s) of scientific maturity" (Bottomore and Nisbet 1979: xv) or is it a "pluralistic, multi-level, reflexive discourse about social life articulated in terms of a number of different and . . . discrete traditions" (Smart 1993: 71)? Our specific answers to this question will shape in part how we define the problem of what exactly is a disaster. Consider, for example, how this problem is defined for the 1995 IJMED special issue.

Professor Quarantelli locates the definitional problem in disaster research square within classical sociology. He writes:

> it is time after nearly half a century of fairly extensive empirical disaster research . . . to systematically address the central concept in the field . . . a developing field will founder unless there emerges some rough consensus about its central concepts . . . unless the field of disaster research comes to more agreement about what a disaster is, the area will intellectually stagnate.
>
> (1995e: 223)

Expressed in Professor Quarantelli's counsel is the strong, foundationalist voice of someone who trusts the promises of a scientific discipline. His words sound the themes of Barry Smart's recent characterization of classical sociology.

> We are advised that a standardization of sociological concepts is long overdue and that a scientific discipline must adhere to some shared set of basic concepts referring uniquely and reliably to its domain of study . . . The promised benefits of standardization are . . . accumulative ordering of knowledge and the achievement of a sense of solidarity among sociologists.
>
> (1993: 70)

Thus, classical disaster sociology assumes the untidy character of disaster studies is but a temporary and repairable state, sooner or later to be replaced by the orderly and systematic rule of a definitive definition of its subject matter. A key theme in the classical tradition is the authority of sociologists to legislate a definition of disaster applicable to all times and places (Bauman 1987). As legislators, sociologists are thinkers as such, persons who formulate ideas about society independent of the meanings, hopes, wishes, or beliefs of ordinary people. Remaining in the secure realm of concepts, one is relieved from intimate contact with the subjects the concepts are supposed to stand for. In classical sociology, legislators, their ideas, and an occasional third-party client, are all that is required to know something important about the world.

Defining the question "What is disaster?" in the legislative voice of classical sociology shaped each of the five papers in the IJMED issue. (A sixth paper by Hewitt is a response to the first five, and, as we will note below, a surprising departure). To make this point consider the following five definitions:

Disasters are:

1 "nonroutine events in societies . . . that involve social disruption and physical harm. Among [sic], key defining properties of such events are (1) length of forewarning, (2) magnitude of impact, (3) scope of impact, and (4) duration of impact" (Kreps 1995a: 258).

2 "a state/condition destabilizing the social system that manifests itself in a malfunctioning or disruption of connections and communications between its elements or social units . . . ; partial or total destruction/demolition . . . making it necessary to take extraordinary or emergency countermeasures to reestablish stability" (Porfiriev 1995: 291).
3 "the loss of key standpoints in common sense, and difficulty of understanding reality through ordinary mental frameworks" (Gilbert 1995: 238).
4 "[events that] release repressed anxiety [and constitute a story of the] loss of control of social order" (Horlick-Jones 1995: 305).
5 "an empirical falsification of human action, as a proof of the correctness of human insight into both nature and culture" (Dombrowsky 1995a: 241).

Two observations suggest themselves at this point: there is an obvious lack of consensus regarding what is disaster and a less obvious problem with the limitations of a strictly legislative voice. First, the problem of consensus.

WHAT IS DISASTER? THE NECESSITY FOR DISSENSUS

Should we be surprised by the absence of agreement among scholars interested in defining disaster? Frankly, no. Look around us, anthropologists cannot agree on a definition of culture (Ingold 1994). Political scientists continue to debate the meaning of power (Nicholls 1990). Psychologists cannot agree on a common definition of memory (Schacter 1996). And, finally, sociologists continue their cacophonous debates about what is a community (Gusfield 1975), alienation (Seeman 1983), and, of course, disaster.

Among the lessons to be learned from the last decades of cultural, behavioral, and social science discussions is the impossibility of reducing the disorderly state of our definitions. These disciplines are dispersed, fragmented, and much farther from consensus now than when they first began. A common definition requires a common community. And, like its counterparts in psychology, anthropology, and so on, modern sociology is anything but unified.

At the 1996 annual meeting of the American Sociological Association, for example, over 450 roundtables and sessions offered thousands of insights, concepts, data, and conclusions, all purporting to be sociology. Over thirty sessions alone addressed the topic of gender (ASA Preliminary Program 1996). Consensus amidst this overwhelming diversity is not possible nor, we would add, is it desirable. Consider, for a moment, the argument that unanimity is something we can live without, indeed must do so.

The remarkable variety of the five definitions in the special issue, each recommended by its authors as capturing essential qualities of disaster, does not promise confidence in a soon to be achieved consensus regarding the definition of this key term, but taken together they express the complexity that is disaster. A single definition—or, for that matter, five definitions—would itself be disastrous for the field. A definition is a way of seeing, a strategy for looking. And every way of seeing, as common wisdom reminds us, is also a way of not seeing.

Imagine for a moment that disaster is a large, pitch black wall in a room with no light. Now imagine each of the five legislated definitions is a flashlight pointed at the wall. Illuminated by each flashlight is a small but important circle of what there is to see and thus know about disaster. At the risk of sounding heretical (and running a useful analogy into the ground), there are probably as many potentially good circles of light to be shed on disaster as there are good sociologists.

While we do not believe classical sociology will succeed in standardizing a definition of disaster, we would not by any means abandon the use of legislated definitions, though we would not rely on them exclusively. A legislated definition of disaster is a particularly useful tool if an investigator is interested in abstract human or organizational activities such as destabilized social systems, etiologies of repressed anxiety, nonroutine events resulting in social disruption and physical harm, collective crisis occasions, and so on. In the hands of a creative analyst, a legislated definition slices a historical, situated subject matter into precise pieces ready to be reconfigured into a comparative abstraction. From this vantage point, disasters are good to think about. They are occasions for us to seriously consider a bountiful range of key social, cultural, and psychological processes.

In addition to their theoretical value, legislated definitions of disaster are a necessary part of applied or third-party research. And we think most readers would agree that a sizable proportion of disaster sociology in America is applied social science with a commitment to problems defined by regulatory agencies and local governments. A popular applied question, How will people respond to civil defense warning systems? for example, begs the use of legislated definitions. In short, this particular definitional strategy is a necessary part of a sociology of disaster and suggests the possible interdependencies between theoretical and applied social science.

LIMITATIONS OF THE LEGISLATIVE VOICE

A second observation regards the limitations of an exclusive use of the legislative voice to define disaster. Defining the problem of defining disaster using the authority of a legislative voice assures each definition is constructed in the absence of local, personal, subjective experiences of disaster, as if sociology and society are two separate enterprises. To illustrate a limitation of this approach, consider an excerpt from an interview the first author conducted with a young man who lived in a small town in Wyoming contaminated by methane plumes expanding and contracting under commercial and residential properties. The question prompting this response did not include the word disaster. The young man was asked, simply, "What does your family think about these methane plumes under the town?"

> My dad says this is a disaster. He goes on and on about his business losing money and the government doing not much to help . . . Mom? She thinks it's a disaster but not like dad does. She thinks we could all end up dead or something of one of those plumes exploding . . . [And you?] I don't know. It seems to me like a flood or earthquake like in San Francisco is a disaster . . . I did say yesterday or some time that if my fiancée moved because of the gases, it would be a big disaster. I don't have a car right now.

If these remarks by a nineteen-year-old who lives with his family atop a methane plume mean something to us as sociologists they do so through frameworks other than those proposed in the above five definitions. Scope or duration of impact, state of social system destabilization, difficulty understanding reality through ordinary frameworks, releasing repressed anxiety, or empirical falsification of human action are not adequate conceptual tools for making sociological sense of his thinking about disaster. It is not, however, that his use of disaster is so simplistic to be below a level of genuine concern to sociologists. For him, disaster is a complicated concept, expressing generational and gender cleavages and also serving as a metaphor to communicate his strong feelings about his fiancée's possible relocation at a time when he does not have a car.

We imagine a reader might object at this point. After all, this young man is not referring to a "real" disaster. Methane plumes were under the town, but there were no explosions prior to the interview. Whether or not a "real" disaster had occurred, however, does not matter very much if our interest is in how this individual (and others in town) experienced and thought about the methane plumes. In the absence of a legislated definition of "real" disaster, we are more apt to see disaster as a key concept organizing the way his mother and father understood the hazard. We are also likely to see that disaster was used to organize one set of experiences for his father and another set of experiences for his mother. And, finally, we can glimpse the rhetorical quality of the word if we recall that while he was not inclined to consider a methane plume a real disaster, like an earthquake, his fiancée relocating to another town would be "a big disaster."

The noticeable gap between legislated definitions of disaster and the situated, local definitions of a young man and his family living in Wyoming illustrate the limitations of this top-down approach. Truth stops at the border of a legislated definition. If we agree to define disaster as this or that, then beyond our definition are only pedestrian and parochial matters, perhaps joined to examples of faulty reasoning among non-sociologists. Indeed, legislated definitions of classical disaster sociology risk being overtaken by the pedestrian and parochial experiences of non-sociologists, an opinion we will defend in the next section.

We must also mention that pedestrian, parochial, and popular definitions of disaster are likely to have far from trivial consequences. Consider a study Aronoff and Gunter (1992) conducted of a non-metropolitan Michigan community that, beginning in the mid-1970s, was discovered to have extensive chemical contamination. Within a few years of this discovery, the county landfill, located on the outskirts of the community, had been placed on Superfund's National Priorities List. The chemical plant that was the initial source of the contaminants, located on a 40-acre site just inside the city limits, was dismantled. In their effort to facilitate recovery from the devastating blow of these contamination problems on the local economy, local officials maneuvered through an array of governmental agencies whose assessments of local circumstances were far from convergent.

From the perspective of the state Department of Natural Resources, which mandated the dismantling of the plant, the contamination posed such an imminent disaster that it required immediate and severe action. From the perspective of local officials, residents, and workers, who mobilized to save the plant, the only "disaster" in evidence was an economic one. From the perspective of the federal Economic Development Administration (EDA), from whom the local government sought a $2 million redevelopment grant, area contamination was regarded as such a significant problem that the EDA would only consider awarding the grant if it could be empirically demonstrated that there was uncontaminated property available in the community. On the other hand, when the local community, working in conjunction with state officials, twice submitted a request to the Federal Emergency Management Agency (FEMA) for monies for disaster relief, the FEMA on both occasions denied the request, arguing that the local community was beset by an economic crisis not a disaster. This small community was thus caught between the pincers of two different federal agencies, one saying they feared the community was far too contaminated to receive money, the other saying the community was not nearly contaminated enough to receive money.

The effort to legislate a definition of disaster may in part be motivated by a desire to move beyond this political messiness. Determining when a disaster can be said to occur should not be left in the hands of those with vested local interests, but rather determined by objective criteria. While the effort to identify the means of distributing public resources other than clout and connections is laudable, we will be remiss if we do not think critically about the political and economic implications of stakeholder groups who seek to define a real world event as a disaster, or conversely seek to resist this definition. Suggested in this example is the need to complement the voice of the legislator with a quite different voice, that of the interpreter. Insuring a space for the interpretive voice is, we will argue, an important step toward opening the sociology of disaster to important new developments. From an interpretive stance the very question "What is disaster?" begs to be reformulated.

HOW IS DISASTER . . . ? THE IMPORTANCE OF THE INTERPRETIVE STANCE

The idea of an interpretive stance in sociology is not new. Symbolic interactionism, ethnomethodology, and the ethnography of speaking are among the more prominent interpretive responses to the hegemony of classical sociology. And Weber, whose work contributed so much to the classical stance, developed his own version of interpretive sociology, *Verstehen*, signaling his ambivalence toward the sociologist as simply legislator.

Recently, however, the interpretive stance has undergone an important self-examination. Gone is the Enlightenment belief that an objective social world awaits discovery through proper application of the scientific method. The best a sociologist can do is capture the conditional, contextual worlds of human beings whose lives intersect at moments to create what we call society. Accompanying this critique of the prevailing charisma of objectivity is an acknowledgment of the profound implications of sociologists in the lives of the people they study. Contrary to what a standard methods book teaches us, we are deeply implicated in our subject matter. We will consider each of these points in some detail, beginning with the importance of the mundane for an interpretive sociology.

The interpretive voice and the importance of the mundane

Interpretive sociology is being stripped of all its pretensions to a naturalistic mode of inquiry. Abandoning its allegiance to realism, it is embracing a vision of the mundane as fluid, multilevel, and composed of a dazzling diversity of people, symbols, issues, and so on. If Blumer and the early Chicago School could write as if there was an empirical world to be revealed through careful observation and concept formation (Clough 1988), Zygmunt Bauman's world is brazenly disordered,

> Whenever one descends from the relatively secure realm of concepts to the description of any concrete object the concepts are supposed to

> stand for—one finds merely a fluid collection of men and women acting at cross-purposes [and] fraught with inner controversy.
> (1993: 44)

And Denzin's recent revision of symbolic interactionism moves smartly away from a realist epistemology, stressing his belief that sociology's "central problem becomes the examination of how interacting individuals connect their lived experiences to the cultural representations of those experiences" (1992: 74). Garfinkel's ethnomethodology constituted a break with Parson's action theory, with its focus on social stability and value consensus. While Parson's worked from the assumption that social actors fairly automatically and unproblematically follow rules and norms, and know which rules and norms apply in various circumstances, ethnomethodologists consider these to be highly problematic and contingent undertakings (Coulon 1995).

This concern with how humans actually "go about getting the job of social interaction done" has resulted in extensive (though not exclusive) ethnomethodological focus on everyday life (Douglas 1970). Anthony Giddens, who is influenced by the perspective of ethnomethodology (Craib 1992), has made humans' interpretive and mundane actions a central focus of his "structuration theory" (Giddens 1984). For Giddens, society is populated by human agents who know "virtually all the time, under some description, what [they are] engaged in and why" (1987: 5).

The mundane is protean. It is, as Denzin reminds us, society with a little "s" (1992: 22). Human beings are agreeing and disagreeing, seeking to know their conditions and circumstances, and moving between questions encountered in their local worlds to more abstract representations of themselves and their habitats formed by newspapers, magazines, televisions, and experts of all kinds, including sociologists.

If society is ordinary people struggling to make a living, committing to or abandoning relationships, responding to untoward events, and so on, then it is appropriate to shift attention from what disasters are to how people create and respond to them. There can be no adequate attempt to explore what disaster is that is not centrally concerned with what it has been said to be. Interpretive strategies explicitly avoid legislating definitions, preferring instead to listen and talk to people who are forever agreeing and disagreeing while inventing and borrowing words and concepts to constitute coherent stories about themselves, others, and the worlds they live in.

An illustration of this interpretive approach may be seen in Aronoff and Gunter's (1992) previously discussed study of a non-metropolitan Michigan community impacted by (what some claimed) were significant contamination problems. That divergent views existed as to whether or not this contamination constituted "a disaster" illustrates a basic contention of the interpretive approach: material aspects of the world do not unproblematically present themselves as instances of this abstract category or conceptual label. Rather, shared collective definitions and designations, if they are even arrived at, are developed through an interpretive and negotiative process.

In their study, Aronoff and Gunter used Giddens' structuration theory to examine the process whereby the local community constructed their own meanings of the contamination in the face of contending claims from outsiders. What is interesting about this case is that the contending claims were exactly the opposite of what has generally been reported in contaminated communities (Freudenburg and Pastor 1992; Brown and Ferguson 1995; Cable and Cable 1995). In this case, it was state and federal governments who were most strongly pushing the "impending disaster" definition, and local residents and officials most stridently resisting this definition. The one exception to this general pattern, where local officials sought a disaster declaration for the pragmatic purposes of gaining federal monies, was the one instance where one federal agency (FEMA) diverged from other government agencies by ruling local contamination problems in fact did not constitute a disaster.

As this study illustrates, interpretive approaches do not start with an a priori definition of disaster, and then ask whether or not a given instance falls within the parameters of that definition, but rather examines the process whereby real-world individuals and organizations struggle to determine whether or not the events that confront them are indeed "a disaster." In making such determinations, individuals look for cues from a variety of sources, including government agencies, sociologists and other scientists, media, their family and friends, the environment itself, and their assessments of the likely consequences of various interpretive stances. Another example from the first author's research illustrates this point. A woman describes changing her mind about the meaning of a mine fire burning under her northeast Pennsylvania borough:

> If you would have asked me about the fire last month you would have got a different answer. Mine fires are what you live with when you live in a mining town. They got one burning more than fifty years in Shenandoah . . . Two things changed my mind about this fire and now I think its a big danger . . . When the DEQ guy says to my husband "take this gas monitor and put it in your living room and if you hear it ring get out of the house," I got scared. Now I feel like there is going to be a ding! ding! ding! and a disaster everyday. [You said there were two things that changed . . .] Oh yea, the second thing that started me worrying was you moving into town. [Me?] Well, what are you here for if there's nothing to worry about?

For this woman, living with the risks of a mine fire was an accepted, though undesirable, part of living in a coal mining town. What changed her mind and moved her toward living on the brink of disaster were the good intentions of a state agency charged with protecting her life and an assistant professor of sociology who could have no earthly reason for being in town if there was "not something to worry about." In the next section we return to the idea that sociologists are likely to change the way people experience their worlds. For the moment, it is enough to point out it was her particular reading of two social responses to the dangers of a mine fire rather than any biophysical events that caused her to shift her stance toward the hazard. Ironically, the act of protecting her (placing a gas monitor in her living room), joined with an effort to understand what it is like to live with chronic danger (a sociologist moves to town), changed the meaning of the mine fire from something you just learn to live with to a disaster waiting to happen.

From an interpretive stance, it is a mistake to attempt to define what disaster is independently of how it is thought about, talked about, and experienced. It is specifically not a sociologist's task to prescribe the strict or essential meaning of a term or activity, but to observe how "ordinary" language itself fashions and shapes human experiences of life's events and circumstances. From this vantage point, there can be no adequate attempt to adjudicate a definition of disaster; rather, attention is directed to how it is experienced in local, situated worlds. Ordinary people will tell sociologists what disasters are if we listen to them.

Consider two published examples of a sociology of disaster from an interpretive stance. Each example suggests how an interpretive voice is apt to contribute something novel to a sociology of disaster.

Surviving disaster and rethinking nature and community

Kai Erikson's *Everything in Its Path* (1976) is arguably the finest example of the effectiveness of the interpretive voice now in print. It is worth noting, however, that not one of the five substantive articles in the special issue cites Erikson. It is Kenneth Hewitt, a geographer, charged with responding to the five papers who finally introduces this important work toward the end of the issue. The significance of Erikson's study of the disaster at Buffalo Creek, West Virginia is the space he provides for the voices of people who lived to narrate their experiences. By listening to survivors of a flood caused by the collapse of a dam holding over 132 million gallons of water and debris, he is able to ask questions about disaster that normally fall outside the range of classic, legislative sociology. Moreover, through the stories of survivors he is able to both affirm and constructively critique key assumptions of classical disaster sociology. We are limited in this paper to a short discussion of only a couple of the good ideas found in *Everything in Its Path*. Perhaps these cursory remarks, however, will spark additional interest in this work.

Listening closely to the narrative accounts of those who survived the flood, Erikson finds "they have clearly lost much of their confidence in the workings of nature" (1976: 179).

> We all walk the floor when it rains a lot . . . So I am afraid of rain. I am afraid there will be something come down some place. There's always going to be a doubt there, always.
>
> (ibid.)

Note we are not told how many survivors lost their confidence in nature or to what degree. Rather, we are shown the words of a few of the survivors themselves that are said to be representative of a subtext in the organization of the self in the aftermath of disaster. To wit, for the survivors of Buffalo Creek, it is natural for nature to hurt them. Disaster and the relationship of self to nature is here identified as an appropriate research question; and importantly, it is derived from local, biographical experiences of a devastating flood and the words survivors use to explain themselves in relationship to the event. Attending to the voices of survivors, Erikson is also able to caution against generalizing a domain assumption of classical disaster sociology. It would appear that not all disasters are followed by a "stage of euphoria" or a forging of a "community of sufferers" as is typically assumed (ibid.: 200–201).

> And so we got up there on the hill and I looked back and said, 'It might have been best if we'd all gone with it, because I don't see nobody else.
>
> (ibid.: 200)

Their recollections of the immediate aftermath of the flood do not include descriptions of altruistic communities spontaneously forming to rescue neighbors. People are not reminding themselves they are among the living by helping others. Rather, they stand listlessly, questioning why they are still alive. In the words of people recalling their feelings and behaviors immediately after a flood, we encounter a variant of the expected (legislated) scenario and are encouraged to speculate on why this case is different.

A key to understanding why Erikson's study is so different from a study of disaster using a legislated definition is his particular definition of disaster and, we might add, where it occurs in the book. For 252 pages he writes knowingly about disaster without defining it! He waits until the last six pages of the book to offer a definition that rests on a variant of what everybody knows about disaster, albeit few of us could say it as eloquently. "[W]hat is a 'disaster' anyway?" he writes, "In social science usage as well as in everyday speech . . . it is a sharp and furious eruption of some kind that splinters the silence for one terrible moment and then goes away" (1976:

253). Guided by this mundane definition, he reveals the complicated and not always successful achievement of human survival.

Erikson's book is complemented by other interpretive approaches to a sociology of disaster. Curiously, however, these other examples of interpretive work examine human responses to environments contaminated by routine industrial practices or unexpected technological accidents (see, for example, Levine 1982; Vyner 1988; Kroll-Smith and Couch 1990a). Erikson himself refers to the Buffalo Creek flood as an "industrial accident" (personal communication). There is, however, a marked absence of interpretive work among sociologists who examine "natural" disasters. We can only speculate here, but perhaps this is because interest in "natural" disasters originated, in part, from federally funded mandates to study questions important to government and not the issues and concerns of ordinary people. Old habits, we know, can be hard to change. This issue begs consideration beyond a simple truism. For the moment, however, we most move on to a second, and arguably more controversial, development in interpretive sociology, the founding of the reflexive voice.

Interpretive sociology and the reciprocal voice

Sociologist Steve Picou who is studying an Alaskan fishing village severely affected by the *Exxon Valdez* oil spill tells the following story. On his last visit to the village he addresses a citizen's workshop on the aftermath of the spill. In the course of his presentation he describes the village as entering the recovery stage. Shortly after the workshop he is standing on a corner and overhears the following conversation between two people, referred to as persons A and B:

A What's going on at the lodge?
B Dr. Picou's back in town making a presentation.
A Well, did he ride in on his white horse to tell us all about our problems?
B He's talking about stages people go through after disaster.
A Really. Well, what stage are we in?
B Recovery.
A Cool. That sounds good to me.

(Personal communication, Steve Picou 1996)

Illustrated in this short exchange is an example of how the insights and conclusions of sociology may enter the mundane lives of people and organizations they are meant to describe, becoming a part of the way local worlds are constituted. Through Professor Picou's talk to a citizen's group, the word "recovery" entered local conversation in an Alaskan fishing village. Recall the woman we encountered above who changed her perception of a mine fire, in part, because a sociologist moved into her town.

Interpretive sociology is now paying attention to what we might call a reciprocal hermeneutic, or the idea that as we are interpreting them they are interpreting us. Sociologist of science, Michael Mulkay, sounds the reflexive voice in his admission:

> I have come to see sociology's ultimate task, not as that of reporting neutrally the facts about an objective social world, but as that of engaging actively in the world in order to create the possibilities of alternative forms of social life.
>
> (1991: xix)

The reciprocal voice recognizes a truth most sociologists have known to be rather obvious, but regularly ignore. Namely, as we study and interpret other people, they are studying and interpreting us. Sociology "does not stand in a neutral relation to the social world . . . social scientists cannot but be alert to the transformative effects that their concepts and theories might have upon what it is they set out to analyse" (Giddens 1987: 71). Embedded in our own subject matter, we change what we study.

The increasing recognition of a reciprocal voice in interpretive sociology (see also Mulkay 1991; Beck 1992; Denzin 1992) stands in marked contrast to the single hermeneutic in classical disaster sociology. Legislated definitions insulate sociology and the sociologist from the simple observation that ordinary people are a lot like researchers searching for ways to understand and make sense of their lives. Indeed, acknowledging the effects of the research act on a group or community is to risk the charge of "contaminating" the data. Outside the rarefied world of the methods textbook, however, is it unreasonable to assume concept-bearing people are watching, listening, and, perhaps, recording sociologists who are watching, listening, and recording them? In the absence of a reciprocal stance that examines the interrelationships between social science and the people it studies, classical disaster sociology does not "see" an important arena of investigation. A concrete illustration of the reciprocal hermeneutic in disaster sociology is found in the work of the first author (Kroll-Smith and Couch 1990a; Kroll-Smith and Couch 1990b).

Sociological knowledge and constructing neighborhood meetings

Kroll-Smith moved into Centralia, Pennsylvania in 1988 to observe residents responding to the uncertainties of an underground mine fire no one seemed able to abate, much less extinguish. Shortly after arriving and making an effort to meet and talk with people, he noticed what for him, at the time, was a peculiar pattern: people assumed sociological knowledge was relevant to the problems they were having both with one another and the fire. He later attributed this surprising interest in sociology to, in part, the high degree of uncertainty surrounding the problems of the fire. Any knowledge, even sociological, was perceived as potentially useful to residents struggling to adapt to its uncertain and potentially catastrophic dangers.

As he approached individuals and groups requesting an interview or the opportunity to observe a meeting, a tacit principle of social exchange was apparently in play. While no one put the exchange principle in quite these words, in deciding whether or not to cooperate with him he was told, in effect, "I will let you observe my activities and tell you my thoughts and feelings if you provide me with information and insights into my predicament." In this fashion, community groups and leaders imposed on Kroll-Smith the interlaced role of researcher/resource and required him to consider some of his own activities as data and speculate on the implications of this for the project. Side by side his professional sociology was a nascent lay sociology that would constitute a part of the Centralia mine fire disaster. Consider a concrete example of this exchange principle in operation.

Kroll-Smith wanted to observe the Centralia Committee on Human Development (CCHD). He approached the chairman of the newly formed CCHD, a Russian Orthodox priest, and asked permission to attend the group's advisory board meetings. The priest agreed to bring the issue before his board. The board decided he could observe only if he agreed to act as a volunteer consultant. In short, the board proposed to exchange the right to observe for his services as a consultant. He demurred, explaining he could not in good conscience serve as a consultant while observing the group; and perhaps more importantly, he did not see what skills he had that would be of any use to their deliberations. In the priest's words, "This is the condition for your participation. You decide." After talking to his collaborator, Steve Couch, he agreed to consult for the privilege to observe the group.

The CCHD worked incessantly for the first several months to establish an organizational vehicle adequate to give citizens in town a forum for expressing their opinions about the fire. Scheduling public meetings in the town hall, however, attracted a self-selected group of angry residents who regularly shouted at one another, hardly the formula for achieving a consensus. Kroll-Smith was asked to work with the advisory board to construct non-confrontational public meetings and encourage residents to participate.

A descriptive survey conducted before he moved into town indicated Centralians expressed strong ties to their neighborhoods. He gave the board a copy of the survey and its members examined the data for what they might reveal about how to devise a successful public meeting. Board members quickly noted that neighborhood attachments based on ethnic and religious affiliations appeared to supersede attachments to community. A person, for example, was first a resident of this or that neighborhood and Lithuanian,

Irish, Scots, and so on, and only secondarily a Centralian. With this observation in mind, the board developed the simple idea of "neighborhood area meetings." Rather than scheduling a single town meeting, the group decided to schedule several neighborhood meetings to be held simultaneously. The goal of these meetings would be to hear from those people who refused to attend the raucous town meetings.

The meetings attracted over 320 people, many more than the forty to fifty that regularly attended the town hall meetings. Importantly, the neighborhood meetings expressed a collective sentiment that was apparently unable to emerge in traditional community forums. Amongst the friendly faces of neighbors and kin, a majority of people confessed their strong desire to relocate, to leave Centralia. In response to this majority opinion, the project board encouraged the formation of the Centralia Homeowners' Association. This group lobbied for and eventually received federal relocation money to buy homes in Centralia at market prices.

In a sociological survey, Centralians revealed how they attached themselves to the community; and later, through their use of this survey, they changed their history in relationship to the mine fire. Both of these data are reported in *The Real Disaster is Above Ground* (Kroll-Smith and Couch 1990a; see also 1990b). Here, the findings of sociology constituted a part of the world we set out to describe.

To be aware that to know something is to change it is arguably a more important lesson for sociologists who study human crises than for those among us who examine more benign activities. A community working to cope with or overcome a crisis, and a sociologist's efforts to record and interpret this work, creates an interdependency between the objectives of the research and the needs of the community that, in our opinion, should be established as a basic part of the research design. Doing so will open the field of disaster research to new, albeit unconventional, questions and insights.

The interpretive voice, as we have set it out here, might appear to some as a call for an "anything goes" sociology. After all, if definitions are situational and unstable and sociologists constitute a part of what it is they study, is not anything possible? The answer, of course, is "No." Sociologists who commit themselves to an interpretive stance must develop valid research strategies to locate the worlds of groups and communities inside themselves while simultaneously recognizing the inescapable articulation of sociology with social life; and they must accomplish this high wire act in a manner convincing to their colleagues. In short, this stance is not without a discipline.

SOWING A FEW WEEDS IN THE GARDEN: A CONCLUSION

What can sociologists know about disaster and societies? As we have said in this paper, we think we do know a great deal and can know considerably more. To clarify this point further and bring this paper to a close, we want to return to the IJMED special issue and introduce and elaborate briefly on the response article by Hewitt. When we look at the sociology of disaster represented in the substantive papers in the special issue we see five carefully cultivated gardens, imposed orders on the wild and frenzied chaos of disaster. To follow any one of these legislated strategies is to till a narrow, but important, furrow of knowledge about people and calamitous events. Like gardens, however, legislative definitions are precarious, encouraging sociologists to protect their prize from the encroaching definitions of other worthy legislators.

Alongside or among the exquisite gardens of legislative sociologists, we believe we need to plant a few weeds. Recalling Will Wright's provocative notion of "wild knowledge" (1992), we recommend considering hazards, disasters, risks, and so on as mundane vocabularies—vocabularies of motives, if you will—that organize how people think and act toward their worlds. We sorely need, as Professor Hewitt notes, "Testimonies from the field," voices of people and communities who live through, with, and among disasters (1995: 326, 329). Recovering the stories of ordinary people is a step toward telling new and arguably significant stories ourselves. In his presidential address to the Mid-South Sociological Association, Stanford Lyman observed that the future of sociology depends on the capacity of sociologists to "tell better and more plausible stories" (1994: 224). Hewitt would have us begin with the stories of others. We would agree.

It is also possible to consider blending legislative and interpretive strategies. Steve Couch and Kroll-Smith attempted this with some success, combining physical characteristics of hazards and disasters with the interpretive strategies of local groups and communities (Couch and Kroll-Smith 1994). Professors Picou, Gill, and Cohen have recently elaborated on this work in their time-series studies of communities affected by the *Exxon Valdez* oil spill (1997).

A final recommendation: take seriously the injunction that sociological knowledge alters the social worlds we study, not as a limitation on our enterprise, but as an invitation to ask new and timely questions about disaster.

15

GLOBAL CHANGES AND THE DEFINITION OF DISASTER

Anthony Oliver-Smith

INTRODUCTION

The social scientific study of disasters has evolved over roughly seven decades to its current multiple conceptual and thematic foci from a variety of origins, each of which contributed in different ways to the formation of the field. In the last ten years there has been some concern that, partially due to its disparate origins and conceptual and practical diversity, the field suffers from a lack of consensus on the concept of disaster that potentially undermines both the intellectual integrity of disaster studies as well as its research enterprise. In short, the question "What is a disaster?" is not considered to have been satisfactorily answered.

However, conceptual or definitional consensus may not be entirely crucial to the health of the field. For example, in 1952, Alfred Kroeber and Clyde Kluckhohn, two of the most eminent anthropologists of the period, published *Culture: A Critical Review of Concepts and Definitions*, in which they analyzed the distinctive properties and components of 164 different definitions of culture, the core concept of the discipline. Although the number of definitions to which anthropologists currently ascribe has diminished, total consensus on the concept has not been reached. Since then, debates have struggled over the issues of whether culture was primary (Geertz 1973; Sahlins 1976), or secondary to material relations (Harris 1979), whether culture is subject to objective scientific analysis or humanistic interpretation, and whether cross-cultural generalizations are possible or whether culture is idiosyncratic, accessible only in its own terms (Rosaldo 1989).

However, lack of complete conceptual uniformity or consensus has not resulted in intellectual stagnation. Although there are some who emphatically disagree with me (Margolis and Murphy 1995), I do not see the continuing debates surrounding the discipline's core concept as particularly damaging to the integrity of the field or as undermining anthropology's research enterprise. On the contrary, the intense self-examination that anthropology frequently becomes involved in serves to revitalize debate and generate new theoretical, methodological, and research questions.

The fact that I do not share the concern over the lack of consensus either about culture or about disaster does not mean that I do not consider the question "What is a disaster?" to be a significant one. The question is and will continue to be significant because it provides an important stimulus and opportunity to explore the varied dimensions of disaster. Indeed, the stimulus and opportunity come at a very significant, perhaps, crucial period in human-environment relations. My own concern focuses less on the lack of consensus and more on seeing a debate engaged in which conflicts may not be totally resolved, but important issues will be increasingly refined and clarified, new perspectives and problem areas explored, and, most importantly, new challenges in practice confronted.

What is it about disasters that has made it so problematic to reach a consensual definition? One problem is that a disaster is a collectivity of intersecting processes and events, social, environmental, cultural, political, economic, physical, technological, transpiring over varying lengths of time. Disasters are totalizing events. As they unfold, all dimensions of a social structural formation and the totality of its relations with the environment may become involved, affected, and focused, expressing consistency and inconsistency, coherence and contradiction, cooperation and conflict, hegemony and resistance, expressed through the operation of physical, biological, and social systems and their interaction among populations, groups, institutions and practices. Disasters when they occur, bring about the conjunction of linkages in causal chains of such features as natural forces or agents, the intensification of production, population increase, environmental degradation, diminished adaptability and all their sociocultural constructions. Like few other phenomena, the multidimensionality of disasters forces us to confront the many and shifting faces of socially constructed reality(ies). An adequate definition and approach to disaster must be able to encompass this multidimensionality.

A second, and probably somewhat less daunting, difficulty is that the task of defining disaster confronts us with the problem of giving analytical precision to a term that has been and is part of general usage. Disengaging a term from popular parlance or literary usage for scientific purposes is problematic, to say the least. The varied popular and literary uses of the word disaster may reflect certain scientific interests, but often include a whole array of conceptual structures, schema and metaphors that complicate attempts at precision, clarity and, most of all, simplicity. Its wide application often embraces far more than would fall under scientific scrutiny.

DISASTER AS A CONTESTED CONCEPT

A potentially useful formulation to discuss the issue of inconsistent or conflicting definitions of disaster is what Gallie has called "a contested concept" (1955). Gallie maintains that "there are concepts which are essentially contested, concepts the proper use of which inevitably involves endless disputes about their proper uses on the part of their users" (ibid.: 169). Concepts such as "art" or "democracy" are disagreed on by differing parties as to their use and application to particular situations or contexts, each party maintaining the correctness of its interpretation with equally compelling arguments and evidence. In daily life in these situations people often agree to disagree or there is some tacit agreement among family members, for example, that religion or politics will never be discussed at the dinner table.

In questions of science or philosophy, however, discussion continues, but with the recognition that contested concepts present special problems. To qualify as contested, there are a number of conditions which a concept must meet. Gallie's five conditions of essential contestedness are: (1) that the concept must be appraisive in terms of some standard of achievement; (2) that the achievement must be internally complex; (3) that the complexity is variously describable; (4) that the achievement be subject to alteration in changing contexts that are not wholly predictable; and (5) that opposing parties recognize that their own uses of the concept are contested by others against whom their uses must be defended (Gallie 1955: 173).

Clearly, the term disaster is appraisive (condition # 1) in that through it certain conditions involving damage and losses are qualitatively and quantitatively assessed. However, the notion of "achievement" (condition # 1) needs to be somewhat reinterpreted. Disasters are associated with a set of effects brought about by a conjuncture of various social, environmental, and technological processes and conditions. Such effects are "achieved" by virtue of this conjuncture. The ideas of production or causation are sufficiently parallel to achievement to permit their use in this context. There is little question that the production of effects ("achievements") which we call disaster is of an internally complex, multidimensional character (condition # 2). Furthermore, the complexity of processes and conditions is variously describable (condition # 3), constituted as it is in terms of economic, psychological, social, political and other impacts and developed from relativistic viewpoints. The complex array of effects often encompassed by the term disaster is also subject to modification over time (condition # 4), both through intention and often without intention, in ways that frequently cannot be accurately predicted. And finally, in some ways, our current project, constitutes the final condition, that of asserting and defending particular uses of a concept against other interpretations. Therefore, I think a case can be made that disaster is a contested concept.

Is the term "disaster" nonetheless useful despite its contested nature?

According to Gallie, for disputed concepts to be of value, two further and more stringent conditions must be met. The concept must be derived from an original exemplar or model whose authority is recognized by all disputing parties. And further, there must be reasonable cause to assume that continued competition by different usages will result in validation or development of the original exemplar (Gallie 1955: 180). Beyond its etymological roots, discussion of original exemplars of the term disaster in common parlance is pointless. Even the early systematic treatments of disasters frequently used the term without specifically defining it. Are there original exemplars whose authority is recognized by all disputants in the disaster research literature? Probably not, although both Fritz (1968) and Carr (1932) are employed by Kreps (1995a) and Dombrowsky (1995a) respectively in such a fashion. The lack of original exemplars notwithstanding, I believe there is reasonable cause to believe that continued discussion of this contested concept will be productive, if not for validation of a nonexistent original exemplar, then for the development of the field.

CONTESTATION OR RELATIVE CONSENSUS?

If disaster is a contested concept, what then is the nature and substance of this contestation? Rather than attempt to reproduce what others, most notably the editor of this volume, have done in tracing differing approaches to disasters (Quarantelli 1985b), I intend to limit my discussion to the most recent efforts to define disasters that appeared in the 1995 issue of the *International Journal of Mass Emergencies and Disasters* (IJMED). In some sense, from the standpoint of the contributors to the special issue of the IJMED, if not that of their discussant, the contested nature of the concept disaster is much more nuanced than might be expected. Indeed, comparing the definitions and approaches offered by the authors in the IJMED with those typically cited thirty years ago (e.g., Fritz 1961), there is now more emphasis on the role the intrinsic qualities of society play in disaster than on the facts of disruption and devastation. Among the IJMED authors there is considerable discussion on common concerns that, while not resolving the ambiguities that have plagued the term, do make progress within their general framework toward that end. There is some agreement on basic issues with more disagreement on how they are to be weighted or applied. In essence, there is a set of common concerns relating to defining what a disaster is. It is less a wildly disparate set of defining characteristics, producing contradictory understandings of a disaster than a case of varying emphases (and in some cases labels) given the specific issues and the shape of their coherence or internal consistency. In some cases, the concerns of some authors constitute subsets of more general issues developed by other authors. Basically, not every author is working at the same level of analysis, a fact which does not preclude achieving some level of agreement or consensus. Thus, each definition in some sense, constitutes an edifice, constructed of similar materials, but of varying designs, most of which have certain areas of compatibility. There is, however, a relative dependence for case material from the developed nations or from extremely large-scale disasters in the developing world that might account for such compatibility. As Hewitt (1995) points out, the inclusion of material from small- and medium-scale disasters, by far the most common and most associated with conditions of underdevelopment, might add greater variation and other voices to the discourse.

To a greater or lesser degree there are at least six areas of common concern among the contributors. All these areas address basic issues in the definition of disasters. First, while there are some slight variations in terms of the importance given environmental factors among them, all reject what Hewitt has characterized as *the hazards paradigm*, which he sees as particularly tenacious, persistently influencing their approaches even within their rejection (1995: 319–20). There is a fairly clear consensus that definitions focusing on agents from the natural or technological environment divert attention from the fundamentally social nature of disaster and impede generalization and theory building. Second, the contributors uniformly situate the phenomenon and the causation of disaster squarely in society and see disaster as defined by its sociocultural dimensions and expressions. Some of the contributors stress certain dimensions over others. Horlick-Jones and Gilbert see the essential features of disaster in the domains of culture and meaning systems and their communication in society. Kreps and Porfiriev emphasize issues of social disruption and destabilization as crucial in defining disaster. Dombrowsky emphasizes the sociocultural nature of disaster causation (as opposed to agency). Despite these differences, there is fundamental agreement that a disaster is at some basic level a social construction, its essence to be found in the organization of communities, rather than in an environmental phenomenon with destructive or disruptive effects for a society.

The third area of general consensus involves societal–environment interaction. The authors, with, I believe, the lone exception of Gilbert, recognize to one degree or another the importance of the interaction of society with the environment in the sense that a disaster is basically the outcome of the interaction between a society with an environmental feature that is determined largely by the society's material and social structures. Kreps interprets disasters as conjunctions of historical happenings (occurring in the social and physical environment) and social definitions of physical harm and disruption. Porfiriev sees all disasters as "ecological", which I understand to mean, situated in and enacted by factors from the social and physical environment. For Horlick-Jones "disasters arise from the behavior of complex sociotechnical systems and the interaction of these systems with their environments" (1995: 306). The extent to which these authors incorporate an ecological dimension in their analysis is not entirely clear. While situating the cause of disaster clearly in societal adaptation, Dombrowsky more specifically than the other authors, delineates the disastrous potentials in the complex interactions between human social organization, material culture, and the natural world, each possessing its own autodynamic processes with planned or understood effects and unplanned, un-understood and unintended consequences. His sense of societal–environment interaction goes far beyond that of the other authors and frames that interaction in an important way.

The fourth area of general agreement among the authors involves questions of the non-routine nature of disasters. Disasters are portrayed as non-routine (Kreps), destabilizing (Porfiriev), uncertainty and disorder (Gilbert), collapses (Dombrowsky), and betrayals (Horlick-Jones). There is clearly an emphasis on distinguishing disasters from ordinary, everyday realities characterized explicitly and implicitly as possessing a higher degree of predictability. Disasters disrupt routine life, destabilize social structures and adaptations and endanger world-views and systems of meaning. While the stress on the non-routine dimension of disasters is perhaps close to common logic, there seems to be, as well, an almost functionalist assumption of general societal equilibrium prior to disaster onset among all the contributors except Dombrowsky. Hewitt's concern about a "tacit assumption of an unexamined normality" is well-founded (1995: 322).

There is also a general consensus on explanations of origins of definitions of disasters among the authors. Different authors assign responsibility for varying definitions of disasters to disciplinary identity or to differences between the goals of research and practice. Research focusing on behavior may define disaster perhaps differently than research focusing on human–environment relations, for example. We also have definitions that are determined by the practice of organizations involved in specific aspects of disaster relief or reconstruction or by specific destructive effects as opposed to definitions elaborated by the research perspectives of different disciplines. Although there is some criticism, notably from Dombrowsky, of operational definitions that discuss disaster as requiring set responses, most of the authors are comfortable with letting practitioners decide what a disaster is for their purposes, while researchers focus on disaster from a more generalized and theory-oriented perspective. Dombrowsky and Porfiriev explicitly try to link practice and research in a mutual search for understanding the nature of disasters.

There are two other questions, which while not central to all the contributions, are nonetheless implicit in the entire project. The first question goes straight to the heart of the definitional problem. What kinds of phenomena should be included within the rubric of disaster and what kinds should be excluded?

A definition that frames a disaster as a crisis in communicating within a community (Gilbert 1995), throwing into uncertainty modes of interpretation and significance (Horlick-Jones 1995), is implicitly broad enough to include such phenomena as a structural adjustment program, the AIDS epidemic, the Oklahoma City bombing, Three Mile Island, the Watts riots, and the Savings and Loan crisis, along with the 1985 Mexico City Earthquake, Hurricane Andrew and Bhopal. Such approaches focus far more on the psychocultural impacts as the crucial characteristics of widely varying kinds of events/processes defined as disasters, and see an emphasis on issues of material or infrastructural damage as failing to address the essential elements of disasters. The kinds of material destruction that other definitions emphasize are, for these approaches, perhaps triggers of the fundamentally sociopsychological or psychocultural essence of a disaster. These broad, cognitively based kinds of definitions that emphasize both cultural and psychological impacts are fundamentally inclusive approaches. They emphasize the dislocating and disrupting effects on human cognition and culture of a wide variety of phenomena that would include the effects of everything from a level five hurricane, a chemical oil spill, a terrorist attack, an epidemic, or a plant closing.

Other approaches, such as Dombrowsky's, based on L. T. Carr's concept of disasters as a collapse of cultural protections, are almost as inclusive. The frame or context in which disasters (that is, the effects of some event or process) occur is a global set of interacting processes of human society and material culture, each with their own internal autodynamics, with nature and its own autodynamics and self-organizing processes. Since we understand so little about the effects of our actions and about these autodynamic systems, the risk of failure becomes very high. Disasters thus become defined as failures of human systems to understand the interactions of this set of interrelated systems, producing a collapse of cultural protections and a resulting set of effects called a disaster. Since disasters can result from the interaction of virtually all-encompassing social, material and natural systems, resulting in a failure of human culture to protect, Dombrowsky's approach is also extremely inclusive.

Other approaches, attempting to balance social disruption, physical harm and psychological dislocation as defining characteristics, are less inclusive, emphasizing physical impacts, but still incorporating a wide array of events/processes. Kreps explicitly wants to include a wide array of phenomena that "involve social disruption and physical harm . . . keeping the boundaries broad to include environmental, technological and sociopolitical events" (1995a: 260). Therefore, civil strife of various sorts would be included under the rubric of disasters in addition to natural and technological events/processes. In his approach, Porfiriev similarly balances emphases on disruptions in communication, partial or total destruction, and physical and psychological overloads, all of which have social space coordinates. Although not explicitly stated, his definition also allows for the inclusion of civil conflict, sabotage and terrorism as disasters. Both approaches would include the psychocultural and social psychological dimensions emphasized by Gilbert and Horlick-Jones, but neither Kreps nor Porfiriev, however, would hold to definitions that would include such social phenomena as economic crises, plant closings, or perhaps computer/high technology failures unless they occasioned specific forms of destruction or mortality. It would seem then that the trend is toward more inclusion than exclusion. In maintaining a balance between social, psychological, and physical effects as the defining characteristics of a disaster, the crucial criterion for the exclusionary definitions seems to be some level of physical destruction.

Another issue that is implicit in most discussions of definitions of disaster is what might be called the "what-why" question. That is, as several contributors state, disasters are to be defined in terms of what disaster is or what a disaster does rather than why a disaster takes place. Porfiriev rejects the concept of vulnerability as relevant to defining disasters. He asserts that vulnerability is more appropriate for explaining the origin and causes of disaster rather than defining it. If we accept the condition that, in Quarantelli's words, "we should stop confusing antecedent conditions and subsequent consequences with the characteristics of a disaster" (as cited in Porfiriev 1995: 292), then our definitions must be conceived in terms of the behavior of human beings and social structures at a temporally and spatially framed moment. Community perception and response, often in the form of organizational involvement, become the crucial variables for defining a disaster (Dynes 1994a). Disaster is largely a behavioral phenomenon and the focus of the definitional problem is primarily the behavior of persons and groups in a specific context of disruption and/or damage as expressed in individual, group or institutional terms. Such an approach is as well very inclusive in that a wide variety of phenomena may elicit the behaviors associated with disasters, thus enhancing the possibilities of comparison among many classes of events and processes. By the same token, including a wide variety of phenomena under the rubric of disaster may tend to obscure significant distinctions across classes of phenomena (Kroll-Smith and Couch 1991).

In one sense, I have no difficulty with defining disaster in behavioral or social psychological terms and applying it to a broad array of phenomena provided the definitional criteria being used are made explicit and the event/process specificities detailed. However, such a definition provides less a starting point than, perhaps, a mid-point to most of the issues about disaster that I wish to explore, namely what disasters reveal about society in (1) its internal social and economic structure and dynamics in relation to (2) its external social and environmental relations, (3) the nature of its overall adaptation and, finally, (4) how this knowledge can be employed to reduce disaster vulnerability and damage.

In a sense, a disaster is symptomatic of the condition of a society's total adaptational strategy. Humans have often been portrayed as the world's most successful species, able to adapt to virtually all of the environments, from the Arctic to the equatorial, the globe has to offer. However, despite flourishing in numbers and complexity, human systems have never been able to absorb or deflect without harm all destructive forces. Furthermore, our own complex systems generate their own hazards, sometimes simply out of inherent slippage among the myriad of complex elements themselves (Perrow 1984). Disasters, and how well or poorly systems fare in them, are in some sense a gauge of the success or failure of the total adaptation of the community.

To return briefly to an earlier issue, if we separate the "why" questions from the "what" questions, we separate the question of hazard from the question of disaster. To disengage the two questions (or sets of questions) is also to disengage from the discussion the term vulnerability or, in Dombrowsky's view, the conditions that produce the essence of disaster, namely, the collapse of cultural protections. In a sense, separating hazard, or "future disasters" in Kreps' terms, from disaster disengages society from the material world in which both are constituted. The "why" is embedded in all disasters.

DISASTER IN ECOLOGICAL PERSPECTIVE

I consider an ecological perspective to be one of the elements essential to the definition of disaster. In advocating the importance of an ecological perspective, I am not suggesting here a return to an environmental *hazards* approach, situating the origin of disasters in environmental forces (*pace* Hewitt), but rather a finer grained understanding of the relationship between society and environment, emphasizing the interaction and mutual constitution of society and environment (Ingold 1992). The focus is on the effectiveness of societal adaptation to the total environment, including the natural, modified and constructed contexts and processes of which the community is a part.

The rationale for adopting an environmental dimension in our understanding of disasters lies in the fact that human communities and their behaviors are not simply located in environments. The interface between society and environment is not one "of external contact between separate domains" (Ingold 1992: 51). That is, environments are experienced by people, and from the experience emerge the constructs of their culture and social organization. At the same time, as people engage their perceived environments in the quest for sustenance and shelter, they are also realized or objectified in terms of its characteristics and utilities for human populations. The development of a society is also the development of its environment and the resulting relations emerge from the multiple continual processes of exchange through the porous boundaries between them. Societies are saturated throughout with their historical environmental relations and environments are deeply conditioned by the histories of societies. To fail to recognize the links between a society and its environment is to remove a society from the history that produced it (ibid.: 51). Society and environment thus are interpenetrating, mutually constitutive of the same world, a world composed basically of the meanings and actions imputed to the abilities and capabilities of people and the possibilities for action provided by natural, modified and built environments (ibid.: 52).

Disasters occur in societies, not in nature, but societies are in nature themselves and that mutually constitutive relationship is not simply given, but is an active, evolving set of interactive processes. In that sense, disasters do not inhere in societies; they inhere in societal–environmental relations. But societal–environmental relations are not relations between two separate entities, but between two mutually constitutive entities. Environmental features and processes become socially defined and structured just as social elements acquire environmental identities and expressions. Excluding the "why" question of disaster in effect excludes the environment which in interaction with a human population contributes to the constitution of society.

Thus, any understanding or definition of disaster must be grounded in a theoretical approach that is capable of encompassing the web of relations that link society (the organization and relations among individuals and groups), environment (the network of linkages with the physical world in which people and groups are both constituting and constituted), and culture (the values, norms, beliefs, attitudes, knowledge that pertain to that organization and those relations). I define a disaster as:

> a process/event involving the combination of a potentially destructive agent(s) from the natural, modified and/or constructed environment and a population in a socially and economically produced condition of vulnerability, resulting in a perceived disruption of the customary relative satisfactions of individual and social needs for physical survival, social order and meaning.

I claim no innovation in this definition. My intention is to merely add to the discourse emerging since the early 1970s in situating disaster in a context of socially produced vulnerability rather than in environmental forces. This construction is also consistent with the recent appearance of formulations on development and environmental degradation, emphasizing the interrelationships among and mutual constitution of the natural and constructed environment, culture, and society (Schmink and Wood 1987; Peet and Watts 1993; Painter and Durham 1995) and similarly founded approaches to disaster (Kroll-Smith and Couch 1991; Bates and Pelanda 1994; Blaikie, Cannon, Davis, and Wisner 1994; Peacock, Morrow, and Gladwin, in press).

The definition I have adopted suggests, as others have, that a necessary but not sufficient condition for a disaster to occur is the conjuncture of at least two factors: a human population and a potentially destructive agent that is part of the total ecological system, including all natural, modified, and constructed features occupied by the human population. Both elements are embedded in natural and social systems as unfolding processes over time. Societies and destructive agents are processual phenomena. Together they define disaster as a processual phenomenon rather than an event that is isolated and temporally demarcated in exact time frames. However, the conjunction of a human population and a potentially destructive agent do not inevitably produce a disaster. A disaster is made inevitable by the historically produced pattern of vulnerability, evidenced in the location, infrastructure, sociopolitical structure, production patterns, and ideology, that characterize a society. The society's pattern of vulnerability is an essential element of a disaster. The pattern of vulnerability will condition the behavior of individuals and organizations throughout the life history of a disaster far more profoundly than will the physical force of the destructive agent.

The question of time becomes crucial if vulnerability is to be considered essential to the definition of disaster. Time has been a basic component of discussions of disaster for almost half a century. Powell, Rayner, and Finesinger (1952) developed a model of time dimensions in disaster, later adapted and employed by Anthony C. Wallace in his study of the Worcester tornado. The time dimensions or states include steady state, warning, threat, impact, isolation, rescue, rehabilitation, and irreversible change (Wallace 1956b: 7–12). More recently, Drabek's encyclopedic compendium of research employs a temporal sequencing as an organizational spine for his research inventory, including planning, warning, evacuation, and other forms of pre-impact mobilization, post-impact emergency actions, restoration, and reconstruction, among other less temporally determined themes (1986). Both discussions employ and privilege time as crucial in understanding disasters, conceptualizing disasters as events/processes. Both the event dimension and the process dimension of a disaster are encompassed by the appearance of a threat generating a warning as the initiation point, a process of development through a number of conditions and some return to stability/normality through reconstruction, all experienced differently by varying social groups, as the end point. Interestingly, the earlier time-phase model begins with a steady state and the later approach discusses planning, indicating an evolution in conceptualizing disaster. That is, a disaster that is planned for is perceived as inherent in the structural properties of a socioecological system and, therefore, must be responded to regardless of the lack of an explicit warning. Planning for earthquakes or hurricanes, or, for that matter, nuclear accidents, takes place within the framework of a processual understanding of systemic features of the environment and the conditions of societal vulnerability.

In line with Kreps' suggestion that the study of disasters should be informed by a life history methodology (1995a: 258), I would suggest that the life history of a disaster begins prior to the appearance of a specific event-focused agent. Indeed, in certain circumstances disasters become part of the profile of any human system at its first organizational moment in a relatively fixed location or area. From my perspective, then an ecological approach appears to be the most capable of encompassing the causation and production of disasters, their development as social and environmental processes and events, their sociocultural construction, and their implications for the overall sociocultural adaptation and evolution of the community (Bates and Pelanda 1994: 147).

A general ecological approach to disasters is founded on the formulation that a human population and its sociocultural system, including all the social, economic, political and technological systems that people generate in their adaptations (or maladaptations) are seen as an element in a total natural system. The sociocultural system is the primary means by which a human population adapts to its environment, enabling a human community to extract from the environment food, shelter, water, energy, and other necessities, and to confront and reduce to some relative degree the uncertainty and vulnerability experienced in interaction with environmental conditions and forces that threaten the population (Bates and Pelanda 1994: 149). To some extent, the forces and conditions in the built, modified, and/or natural environments that characterize disasters are forms of adaptational challenges to which the society must, but does not always, adequately respond. In so far as it is impossible to guard against every threat completely, all systems experience degrees of inherent vulnerability. Thus, the problem presented by hazards and disasters is framed within the overall pattern of societal adaptation to the total environment.

A POLITICAL ECOLOGY OF DISASTER

However, the danger in thinking about adaptation in general or disasters in particular in terms of general ecology is that the human system is usually formulated as unitary, one element of many interacting in a pattern of energy flows with other elements in the total system. The human population and its intricate sociocultural system becomes a "black box," devoid of internal characteristics, when it is obviously complex and internally differentiated to one degree or another. The complex internal differentiation that characterizes all but the earliest levels of sociocultural integration may distribute the benefits of adaptational effectiveness in widely disparate ways in both short and long terms. From this perspective, the patterns of adaptation developed out of the social systems of the society may be effective generally, effective only for those favored by the societal power relations or patterns of production and allocation, or ineffective for those not so favored. The same patterns of adaptation, while reasonably effective for some or many in the short run, may equally sow the seeds of future vulnerability and disasters in the long run. The complex internal differentiation generally characteristic particularly of contemporary human societies thus requires the combination of an ecological framework with an analytical strategy that can encompass the interaction of environmental features, processes, and resources with the nature, forms, and effects of the patterns of production, allocation, and internal social differentiation of society. The fact that societies, as adaptive systems, are controlled by contesting interests within a society, privileging some sectors with enhanced security, while subjecting others to systemic risks and hazards, is also apprehended by this strategy.

In essence, what is required is a political ecology of disasters. Political ecology situates an ecologically grounded, social scientific perspective within a political economy framework by focusing on the relationships between people, the environment, and the sociopolitical structures that characterize the society of which the people are members (Campbell 1996: 6). A political ecology perspective on disasters focuses on the dynamic relationships between a human population, its socially generated and politically enforced productive and allocative patterns, and its physical environment, all in the formation of patterns of vulnerability and response to disaster. Human–environmental relations are always structured and expressed through social relations that reflect the arrangements by and through which a population extracts a living from its surroundings. Such an approach recognizes that the social institutional arrangements through which human beings access and alter the physical environment in their quest for sustenance and shelter are key elements in the evolution of disasters. The analysis focuses on those conditions surrounding the disaster, either threatened or occurred, which condition and shape its evolution, most particularly those structures that shape the forms of development that make the society vulnerable to both socioeconomically and environmentally generated hazards. A brief summary of an analysis of the 1970 Peruvian earthquake informed by these formulations is offered as an example of such an approach (Oliver-Smith 1994).

THE FIVE-HUNDRED-YEAR EARTHQUAKE

On 31 May 1970 an earthquake registering 7.7 on the Richter scale took place about fifty miles off the coast of Peru near the city of Chimbote. The earthquake affected an area of about 83,000 square kilometers, larger than Belgium and the Netherlands combined, claiming 65,000 lives, injuring 140,000 people, and destroying or damaging more than 160,000 structures, roughly 85 percent of the total building stock. Over 500,000 people were left homeless and approximately 3 million others were affected both on the coast and in the Andean highlands. Economic losses surpassed half a billion (1970) dollars. One hundred and fifty-two provincial cities and towns and over 1,500 peasant villages were seriously damaged or destroyed (AID 1970).

When I began studying this colossal phenomenon in 1970, I basically approached it as a historical geological event that triggered a complex series of interrelated social processes that were to continue at fairly intense levels for close to a decade and, in some cases, continue today in the form of altered practices and policies. However, after the initial research was completed, I became puzzled by the contrast between the extraordinary mortality and destruction in 1970 and the lack of evidence of comparable mortality and damage in the Andean prehistoric record. The Peruvian coast and highlands are characterized by a series of natural forces and phenomena with enormous potential for destructive power when combined with human populations in vulnerable configurations. These two interconnected regions are chronically prone to earthquakes, volcanoes, floods, droughts, and other hazards. As these complex and unstable environments have been home to human inhabitants for over 10,000 years and the site of major cultural complexity for the last 4,000 years, the nature of human cultural adaptation to these environments becomes a compelling issue.

Exploration of the adaptations of pre-Columbian Andean peoples to their hazard-prone environment revealed five basic patterns: (1) control of multiple ecological tiers, thereby diversifying production and spreading risk (2) dispersed settlement patterns, with little nucleation or density among the general population (3) environmentally appropriate building materials and techniques, emphasizing lightweight roof materials and low, thin walls for vernacular architecture (4) preparedness in the form of storehouses called *qollqas* and supplies set aside for the state's use which included responding to local privation, and (5) an ideology expressing a cultural tradition of disaster awareness. Although environmental forces, particularly tectonic uplift (Moseley, Feldman, and Ortloff 1981) and drought (Kolata 1993) appear to have played important roles in large-scale culture change in the Andes as a whole, these adaptations seem to have been relatively effective in enabling Andean peoples not only to survive, but to flourish, if population growth and cultural complexity are any measures of success.

However, for the Andean peoples the conquest of Peru signified a cataclysmic demographic collapse, due largely to indigenous vulnerability to European disease, and the distortion or destruction of the adaptive systems to their environment. In their quest for wealth, the Spanish attempts to control and exploit the large population subverted specific indigenous adaptive strategies to their hazardous environments. The Spanish were both ignorant and largely uncaring about Andean notions of territoriality and settlement patterns. Spanish approaches to settlement location flew in the face of pre-Colombian experience with hazards. They located towns at the confluences of rivers, where they were vulnerable to landslides and floods and near active volcanoes, which in the case of the city of Arequipa resulted in its destruction four times in the seventeenth century alone (Cook 1981:171). In order to control the Andean population, in 1570 it was ordered that Andean communities be concentrated or "reduced" from their dispersed settlements into planned nucleated communities.

Spanish building techniques and settlement design were employed in the *reducciones* (new communities) for Indians and the new towns and cities founded by the Spaniards. Unlike the dispersed pattern of Inca towns in which houses were spaced out along long-used paths, Spanish settlement design favored the grid pattern of perpendicular streets organized around a central plaza. The streets tended to be narrow and the houses adjoining or close together. Many houses in these Spanish towns had a second-story storage area as well, something which few domestic dwellings had in the pre-Colombian era.

While the building materials used in the pre-Colombian era—adobe, stone, and thatch—continued to be used, some dangerous changes in materials appeared. Clearly, the most dangerous of the changes gradually adopted was the ceramic barrel roof tile. Building techniques such as tying walls together at the corners also began to be abandoned, creating the conditions for the construction of a very seismically vulnerable dwelling. Houses with untied walls constructed of adobe bricks built two stories high and topped with an extraordinarily heavy ceramic tile roof are extremely unstable and dangerous in an earthquake.

Changes in settlement patterns compounded the danger such dwellings presented for their inhabitants. The more densely organized Spanish settlements with their narrow, perpendicularly arranged streets lined with one- and two-story houses of adobe and ceramic tile roofs combined to create a very dangerous situation. Narrow streets, untied walls, heavy roofs, and seismic tremors are a deadly combination.

Finally, the important institution of the *qollqas*, the storehouses, was also lost. The assiduous extraction of surpluses by the Spanish precluded, after a very short time, any amassing of stores for contingency purposes. The *qollqas*, so vital in precluding long-term local privation after disaster impact, eventually fell into disuse and general abandonment in the first century of the colony, leaving the decimated and demoralized population even more vulnerable to further catastrophe. The gradual subversion over time of the social and economic systems of Andean communities, based first on the extraction of tribute and later on peonage, usurpation of land, and terms of unequal market exchange in a system dominated by colonial and later national and international interests submerged the region in abject underdevelopment, leaving them in a condition of extreme privation and vulnerability.

Thus, the earthquake of 31 May 1970 devastated a segment of a highly vulnerable region and became the worst natural disaster in the history of the Western hemisphere. The high mortality rates in highland cities, in particular, were due largely to three major factors: settlement location, settlement plan, and building techniques and materials. Avalanches loosed from Andean peaks by the earthquake tremors careened down the canyons of rivers to obliterate several villages and cities located in these natural channels. The worst of these avalanches descended from Mount Huascaran, Peru's tallest mountain (almost 6,800 meters), upon the provincial capital of Yungay, destroying and burying the city and roughly 4,500 of its 5,000 inhabitants (Oliver-Smith 1992).

The second factor, settlement planning, combined with the third factor, building techniques and materials, and the earthquake to convert town and city streets into death traps. When the earthquake hit, the untied exterior walls of buildings, subject to the excessive weight of tile roofs, fell outward into the narrow streets, burying people who attempted to escape, and the heavy roofs fell into the house upon those who remained within. Highland cities and towns became massive death traps for their inhabitants when the earthquake struck. In the departmental capital of Huaraz, almost a third of the population, some 10,000 people, lost their lives in this fashion.

In the final analysis, much of the devastation and misery caused in Peru by the earthquake of 31 May 1970 was a product of the historical processes set in motion at the time of the conquest. These processes ultimately subverted the generally effective adaptations to the many environmental hazards worked out by the peoples and cultures of the Andes over the 10,000 years of human residence in the region. The accentuated vulnerability which the region exhibited and still exhibits is a socially created phenomenon, a historical product brought into being by identifiable forces. The destruction and misery in the aftermath of the 1970 disaster were as much a product of Peru's historic underdevelopment as they were of the earthquake. In effect, the earthquake which devastated the north-central coastal and Andean regions of Peru can be seen as an event/process which in certain respects began almost five hundred years earlier with the conquest and colonization of Peru and its consequent insertion into the developing world economic system of which Spain was a major architect.

CONCLUSION: DEFINING DISASTER IN THE CONTEXT OF GLOBAL CHANGE

I have argued for and attempted to demonstrate the need for a political ecological perspective that discloses the social, political, and economic forces and structures that condition human–environment relations in defining and researching disasters. I base this argument on the contention that human societies and their environments are fundamentally inseparable, engaged in a continual process of mutual constitution and expression. Since the structures through which human–environment relations are expressed are fundamentally social in nature, they are characterized and largely reproduced by the dominant forms of production that generate major forms of differentiation and reflect the patterns of allocation in the society. As disasters are expressions of the interface of human–environment relations, an approach is required that can encompass both the environmental and the political economic dimensions in a single analytical format focused on adaptation.

The need for such a conceptualization and approach is becoming more urgent as the nature of hazards and disasters rapidly evolves. The conceptual scope of adaptation has broadened recently to include not only the issues of success or failure of species within environments, but the viability of the environment itself as a self-sustaining system. From the human perspective, the question of how well a society is adapted to its environment must now be linked to the question of how well an environment fares around a society. The issue of mutuality is now at the forefront. Disasters more than ever now express most clearly imbalances in that mutuality.

We are now realizing that there are limits both to human adaptive capabilities and to the resilience of nature and, further, that the continued expansion of human activities in the world is straining those limits. Currently, a spectrum of problems is emerging, caused by human effects on air, land, and water that slowly gather momentum until they trigger rapid alterations in local systems that impact the health of populations, the renewability of resources and well-being of communities. Furthermore, with the increasing globalization of biophysical phenomena coupled with a similar globalization of trade and migration, a process of intensification of linkages is creating problems across scales in space and in time. In effect, local problems today may have their root causes and triggering agents, and possibly their solutions, on the other side of the globe. This globalization process means that problems are basically nonlinear in causation and discontinuous in both space and time, and therefore, inherently unpredictable and fundamentally precluding of the traditional response of observing a signal of change and then adapting to it. These problems place both societies and natural systems in such basically new and unknown terrain that both their social and the ecological elements have evolutionary implications. Basically, people, economies, and nature are now in a process of co-evolution on a global scale, each influencing the others in unfamiliar ways and at scales that challenge our traditional understandings of structure and organization with serious implications for the adaptive capacities of people and societies (Holling 1994: 79–81).

The implications of these conclusions for the study of disasters are profound. They emphasize that the nature of disasters is rooted in the co-evolutionary relationship between human societies and natural systems, and oblige us to intensify our efforts to specify the linkages, now on regional and global scales, that generate these destructive forces within our societies and environments. Disasters are now becoming sentinel events of processes that are intensifying on a planetary scale. Our definitions and our approaches to study them must now reflect these realities.

SECOND REACTION ARTICLE

16

DEFINITIONS AND THE DEVELOPMENT OF A THEORETICAL SUPERSTRUCTURE FOR DISASTER RESEARCH

Ronald W. Perry

The challenge given contributors to this volume was remarkably straightforward: define a disaster. Yet the task is deceptively complex. For a social scientist, offering a definition is the outcome of a process shaped by the person's philosophy of science, their training and experience, and the purpose to which the definition will be put, among other factors. Relative to purpose, I see the goal here broadly, aimed at placing some sort of parameters on what has become a field of study; where "disaster" has been alternately treated as a "cause," an "outcome" or a "context." As Quarantelli (1995b) has pointed out, the growth of disaster research as an endeavor is dependent upon the development of a *theoretical superstructure* that is tied to at least some consensus regarding the phenomenon. In a broad theoretical sense, I see the comparative examination of proposed definitions as one barometer of consensus.

Each author in this section has presented not only a definition, but an elaboration of disaster phenomena. The task of the discussant is to identify and review the definitions, comment on their implications and degree of consistency, and draw conclusions about issues they raise for the field. Since my assigned task was to comment only on these authors, I do not critique or explicitly address, except in passing reference, the proposed definitions from the first section which were originally published in a special issue of the *International Journal of Mass Emergencies and Disasters*. I shall address the discussant's tasks in three sections. First, I will describe each definition and elaboration with only minor contextualizing of each author's views. Second, I will offer commentary and critique of each definition. Finally, the paper closes with discussion of a series of issues that are important to the further development of the field.

THE DEFINITIONS

Russell Dynes indicates that "a disaster is a normatively defined occasion in a community in which extraordinary efforts are taken to protect and benefit some social resource whose existence is perceived as threatened." Thus, disasters are tied to the social unit of community, which may be (geographically) a place, but one that has a social psychological component of identity. Within this socially defined context, disasters are "occasions" (perhaps an opportunity for action in the span of social time) that provoke action to preserve a social resource believed endangered. In terms of various time and stage and phase continua (originally Carr 1932; or National Governors' Association 1979), "disasters" are understood in the "emergency period" or response phase. This is important because extraordinary effort is characterized or measured in the levels of involvement and the types of community organizations activated to protect the resource.

This definition is inclusive and broad. To further specify meaning, Dynes offers the rudiments of a typology or simple classification system that in effect extends the definition by distinguishing different types that constitute subcategories of disasters. The scheme offers three general types of disasters, identified by the presence of the community (as the location of social action) and the presence (or use) of local versus external organizations in mounting extraordinary efforts. Only three of the four (local–external by community–non-community) possible types are elaborated. *Autonomous* disasters are those that impact the community and are handled with local organizations. *Dependent* disasters are community disasters managed with external organizations (with the implication of some local involvement). *Non-community* disaster is used to capture any situation where the social unit is not the community whether the response is based on local or external organizations. In specifying the meaning of disasters further, Dynes identifies subtypes for each of the general types.

With elegant parsimony, Robert Stallings indicates that "disasters are fundamentally disruptions of routines." The context for the definition is the notion of the ongoing operations of societies or human groupings. Structures of norms, beliefs, practices, and so on evolve that constitute routines for dealing with the needs and demands of individual and collective existence. These become the routines of life, or possibly just life (social psychologically speaking). Over time, changes (internal in origin or externally imposed) in the setting or environment may render a given routine either ineffective or impossible to implement. For Stallings, I believe these kinds of changes represent "disaster," although he clearly indicates that disaster is only one kind of many kinds of exception. In extending his conceptualization of disaster, Stallings observes that disruptions evoke responses in the form of exceptions to routines. He notes that "exception routines" represent either established (but "stored" or possibly "normally" used in a different context) or emergent patterns for temporarily achieving the desired end. Over time, faced with continued success of the exception routine, one could expect that it would become an institutionalized exception, or that it could form the basis for reformulating the approach to the need itself.

For Anthony Oliver-Smith disaster is "a process/event involving the combination of a potentially destructive agent from the natural, modified and/or constructed environment and a population in a social and economically produced condition of vulnerability, resulting in a perceived disruption of the customary relative satisfactions of individual and social needs for physical survival, social order and meaning." In elaborating, Oliver-Smith emphasizes the importance of an ecological perspective, acknowledging the "interaction and mutual construction of society and environment," and "situating disaster in a context of socially produced vulnerability." Hence it is not environmental phenomena per se that are key to disasters, but the interaction of the human use system with nature and the prospect of vulnerability on the human side (perhaps limits on the ability to cope). Vulnerability is socially produced, but may be related to the state of technology (either the extent of technology or its efficacy).

In discussing the causes, effects, and responses to disasters, Oliver-Smith creates a framework for the political ecology of disasters. As was the case with Dynes (who used taxonomic reasoning to produce a classification), this elaboration serves to further explicate disaster and to identify corollary concerns of antecedents, characteristics, and consequences without confusing them in the original definition itself. Social change is an explicit concern here. The focus is placed on understanding human behavior and institutions in their natural and constructed environment—trying to deal with aberrations stemming from situations where personal, institutional, and cultural mechanisms that usually insure smooth social intercourse fail. In this sense, disaster is an occasion or opportunity in social time to protectively respond in the short-term and to "adapt" in the long-term. By extension, successful or at least "consensually satisfying" adaptation (through social, cultural, or technological solutions) to recurrent threatening events/processes removes that disaster. Oliver-Smith acknowledges, through his political economy framework, that one can understand (and expect) differential impacts and both access to and effectiveness of adaptations will vary by individual membership in different societal subgroups.

Uriel Rosenthal approaches the notion of defining disasters from a developmental perspective, through the device of contrasting traditional conceptions with contemporary conceptions. He identifies the traditional view of disasters as one which largely focused upon sudden onset, narrow scope impact, natural events, taking place in a "clearly demarcated context." And certainly in my experience, for many years events with these characteristics occupied most of the fieldwork attention of disaster researchers. In explaining the meaning of the term disaster, Rosenthal takes two tactics.

First, he embraces Dynes' (1974) notion of *compound disasters* and by implication accepts the rudiments of Dynes' original definition (which reflects the definition presented here). Rosenthal then personalizes the definition by specifying aspects of contemporary disasters and in so doing indicates aspects of the definition that require rethinking to accommodate contemporary events and perspectives. In characterizing disasters, Rosenthal emphasizes "linkages of conditions, characteristics and consequences." Indeed, "connectedness" and "process" appear to be key parts of his view of disasters. Thus, he does not so much exclude the traditional approach as he argues that one needs to think about its elements differently. For example, Rosenthal does not explicitly dismiss the notion of disaster as an *event*; he does say that the notion of disaster event needs to be thought of in a different way. Namely (and similar to Dynes), as an occasion with a potentially elaborate network of "causes" rooted in the past, with a similarly complex and far-reaching constellation of "consequences" that are contemporary, and with "ripple effects" that proceed forward in time.

For Rosenthal, a given socially defined "disaster" constitutes one focal event among many potentially related events over time and throughout the social structure. In this context, the particular disaster agent is not particularly critical; disasters are socially defined and the social definition may even vary between actors, alerting researchers to the question of "disaster for whom?" Change is also an inherent part of Rosenthal's formulation. He mentions "creeping disasters" to emphasize both the importance of gradual onset events and the point that future disasters may have this type etiology more often than in the past. His notion of change following a disaster is not a "steady state" return, a vestigial remain of classic systems thinking, but a call to more broadly examine and identify social system accommodations and innovations.

The approach by Kroll-Smith/Gunter to the definitional task is somewhat different from that of the other authors, concentrating at least as much on philosophical questions as a substantive definition. They begin by acknowledging (and endorsing) that there can be many answers to the question of "What is a disaster?" On the one hand, they report seeing less consensus about the definition of disasters than the other authors, but seem to agree that "dissensus" is not a debilitating problem. Much of the paper is spent in critique of what Kroll-Smith/Gunter see as a problematic "classical model" of disaster research. Without detailed specification of either approach epistemologically, they contrast the classical model as based principally on what could be called positivist foundations with an interpretative approach apparently based more on phenomenological principles.

Building on the idea of what constitutes a definition, Kroll-Smith/Gunter acknowledge that "legislated definitions" (like those proposed by the other authors) do exist and have a place in theory construction and applied work. They then contend that such definitions are "constructed in the absence of local, personal, subjective experiences of disaster, as if sociology and society are two separate enterprises." Much of their discussion is devoted to advocating an "interpretative stance" that presumably incorporates the more subjective side of the experience of the phenomenon being studied. Kroll-Smith/Gunter offer a variety of examples of the subjective view of disasters from his research and that of others. Clearly, the quotes offer a variety of rich personal experiences and perceptions, from which one might inductively begin the process of identifying dimensions that could develop into a concept of disaster.

Kroll-Smith/Gunter do not offer a "legislated definition" of disaster. At one level, it appears that the definition which grows from an "interpretative stance" is that "disasters are what people say they are." In places they refer to disaster as an "occasion" (similar to the other authors), but seem to be thinking of disasters more as a "cause" of behaviors than a "context" in which behaviors take place. It appears that the "event"—its physical character—is important to the extent that it shapes the way people conceive of it and react to it. A quick content analysis of the examples and interviews recounted by Kroll-Smith/Gunter might shed light on what they believe a disaster is, but it would be their vision and not those of others. Also, I think the tactic of not formally offering a definition is part of their point in writing the paper.

While much time is given to the delineation of problems and limitations of "classical model" and "legislated definitions," Kroll-Smith/Gunter do say that these are a necessary part of a sociology of disaster and suggest the possible interdependencies between theoretical and applied social science. They thus seem to be arguing that the problem lies not in the existence of legislated definitions, but in depending and focusing exclusively on such definitions, apparently to the exclusion of insights from the "interpretative stance."

COMMENTARY

There is much consistency in the approach to the problem and the definitional substance in the work of Dynes, Stallings, Oliver-Smith, and Rosenthal. Kroll-Smith/Gunter offer advice on the meaning and construction of definitions and recount numerous reactions to disaster events from people experiencing them, without themselves offering an explicit definition. In this discussion, I wish to comment on the compatibility of the definitions offered by the authors who proposed them. Also, each author discussed a variety of issues that relate to or expand upon definitional matters; many of these dealt with the future growth of disaster research. I shall use this section to highlight and explore some of the implications of suggestions that I feel are particularly important.

Dynes, Stallings, Oliver-Smith and Rosenthal each seem to agree on a number of key definitional dimensions of disasters. Disasters are socially defined events in social time. Disasters create disruption of social intercourse and, in that regard, are a context for social action, but the disaster agent itself is not a key component of the definitional task. The events, occasions, or processes labeled disasters can be part of an external environment (*nature*) or internal in the sense of social (technological or even human intervention) manufacture. Disasters must be understood in relation to social change: as occasions that offer an opportunity for (or perhaps actually *demand*) creation and adoption of short-term correctives or long-term adaptations.

Although stated with different words and different emphases and elaborations, the consensus evidenced here is unmistakable. Dynes and Oliver-Smith pack many of the above elements directly into their definitional statements. Stallings' definition is shorter, but his elaborations cover most of the same elements. Rosenthal presents his definitional features through comparisons of traditional and contemporary research-practice driven definitions and through examples. I believe that each author's "vision" of a disaster—inferred from his definition and discussion—is ultimately very similar to the visions of the others. It is also worth mentioning that the consistency in *vision* that I see here extends, as Stallings pointed out, to aspects of the definitions proposed by Kreps, Porfiriev, Gilbert, and Dombrowsky. That is, although I think the definitions from the journal issue show less consistency with each other than Dynes, Stallings, Oliver-Smith, and Rosenthal, one finds many common *elements* across the entire group. Furthermore, there is a basic consistency between all these definitions and many of those classic to disaster research beginning with Prince (1920), Carr (1932), A. Wallace (1956a) and Fritz (1961). Certainly differences can be found. Perhaps because of its appearance in a social problems text-reader, Fritz's definition is sometimes identified as the beginning of a "social problems" view of disaster, rooted in functionalist tradition, as presumably an alternative to social change or social constructionist perspectives (cf. Kreps and Drabek 1996). However, a common thread is discernible; some differences can be understood as arising from inductively stimulated reconceptualization over time, while other variations reflect different disciplinary or theoretical frameworks, or unit of analysis orientations of the social scientists.

In dealing with nominal definitions (Hempel 1952: 3), there are many possible tests for cross definitional compatibility. What I have done here is to look for three (admittedly arbitrary) elements: (1) a statement of how disaster is recognized—Is it agent-based or socially defined? (2) a statement of its status as cause, effect or context for behavior; and (3) a statement of its relationship to social time and social structure. I see these elements as fundamental definitional features, realizing that there is almost inherent variation in definitions related to the multidimensionality of concrete situations (Blalock 1982: 110). Any author's definition may also identify or discuss antecedent conditions for disasters, consequences of disasters, or characteristics (features) of disaster agents. These latter features are a matter of elaboration and likely to be more subject to variation between researchers than the key elements because such elaborations are likely to reflect authors' different theoretical emphases.

In developing a definition for disaster, each author also raises important issues relative to the creation of the theoretical superstructure mentioned by Quarantelli. Dynes provided a general definition and further explicated it by creating a simple classification of types of disasters. In one way, the typology is an extension of the definition, identifying and collecting the particular elements of several (the categories and subcategories of the classification) different kinds of disasters. Relative to the future direction of research and theory, Dynes is alerting scholars to expect differences between the disaster types and develop explanations within them. Of particular theoretical interest are two of the features used to create the classification.

Dynes explicitly uses the community as the significant social unit. While the community has long been the apparent focus of disaster thinking, most empirical work has dealt with (physically and systemically) subcommunity units. Although one could argue that a "significant subsystem" of a community is actually understood only in view of the whole community, few among many "disasters" studied have been "community-wide." Fritz's classic definition seems almost crafted to account for this condition.

Using the notion of community in a definition leaves one with a rather large range of events that might be called disasters *except* for their involvement of social units other than communities. Dynes handles this matter within the classification by including *non-community* disasters. He accurately notes that as constructed, this category could exactly reflect the classes and subclasses of (autonomous and dependent) *community* disasters. Technically (when crafting typologies) this is a workable solution, although Dynes reports it to be not intellectually satisfying. I suspect the discomfort stems from the conjunction of two matters. First, it is clear that Dynes wants (like many if not most other disaster researchers) to separate what Quarantelli has called "accidents"—events that are dealt with routinely by "constituted" emergency responders such as house fires and the like (cf. Perry 1990)—from the category of disasters, and incorporating a community-wide feature into the thinking is one effective way of accomplishing the exclusion. Second, however, Dynes appreciates that the community is one unit of analysis, among many others available (and of theoretical interest) to disaster researchers, and in that sense the exclusion is not so satisfying.

At one level of methodological discourse, the next step for Dynes' classification would be to explicitly name "non-community" units of analysis and expand the classification to accommodate them. Although intricate, this is a finite and executable step that would have the desirable effect of providing an (albeit massive) framework for research and theorizing. A different option would be to leave the typology alone, emphasizing its utility for creating categories for types of community disasters and eliminate the non-community categories. This would simply become a narrower typology. Still a third option is to keep the logic Dynes develops for definition, but substitute the term "social system" for community in the definition. This changes the nature of the task and "generalizes" or makes the goal more generic. Dynes could still link the term "disaster" exclusively to community units. The thrust of the typology, however, would focus on some more generic kind of "system stress" or "uncertainty situations" in social systems with "disasters" as a type that would clearly spawn many (as he has already described) subtypes. This follows the logic that Barton (1969) used in developing a typology of "collective stress," but Dynes' version would deal more particularly with "disasters."

Another notable feature of Dynes' work is the tactic of distinguishing *types* of disasters based in part on the "types" of organizations activated during the emergency period. There are at least two impressive aspects of this decision. First, it cleanly captures what are historically thought to be important dimensions of disasters: especially the involvement of collective resources. Second, it demonstrates the fecundity of typologies. Dynes began with the fourfold DRC typology of organizations; he mentions the opportunity for greater precision by adopting the 64-class DTRA typology created by Gary Kreps (1989a). In either case, a typology becomes the basis for classifying disasters (creating another typology) in terms of the external–internal activation and community versus non-community dimension. Dynes' practice is both consistent with the argument that typologies represent an important avenue for theoretical growth (Kreps 1989c) and demonstrates the utility of "nesting" typologies (Perry 1989a).

Robert Stallings defines disaster, also discusses taxonomic options, and creates a compelling logic linking disaster research with the classic problem (and theory) of social order. His elaboration focuses upon the need to understand human behavior using the techniques and tools of sociology (also other disciplines studying social order). Strategically these arguments suggest growth does not stem from the creation of a "theory of disaster" or necessarily from "middle range" theories of disaster phenomena. Instead, one would conduct research and develop theories of social processes looking at disasters as one context for social behavior, perhaps under the rubric of uncertainty. And, in explicating routines, exceptions and exception routines, Stallings creates conceptual tools and theoretical rationale that demonstrate the connection between disaster and non-disaster contexts permitting the use of (and perhaps extending) traditional sociological concepts and theories.

There is special intellectual appeal to the tact adopted by Stallings. First, it provides for and encourages continuity in the discipline. Certainly phenomena arise in disaster contexts that demand new concepts and independent explanations. However, much theorizing about the social order (the "core questions" of sociology) already implicitly or explicitly deals with departures from order which subsume the uncertainty found in disasters.

Stallings reminds readers that many disaster research pioneers saw disaster settings as opportunities to examine implications of concepts, theories, and rationales for understanding basic social processes. Stallings clearly and impressively reminds us of Durkheim's prescription to study "social facts."

A second point of appeal associated with Stallings' definition and approach is that it comfortably fits many different units of analysis. While he is personally more interested in a macro social level (and discusses disasters largely in these terms), his parsimonious statement of routines, exceptions, and exception routines fits elsewhere too. Perhaps much strength stems from Stallings' avoidance of the pitfalls of macro social theorizing outlined by Gerhard Lenski (1988). Through careful attention to conceptualization and consistent use of precise terminology he creates testable, linked claims. Ultimately, I believe that the framework is *facilitating* rather than confining. Routines, exception experiences, and exception routines can be applied when thinking of individuals, families, organizations, communities, or societies (albeit with some distortion of Stallings' original intent). Indeed, there is an interesting parallel between Stallings' approach and the emergent norm thinking of Ralph Turner (1964), originally aimed at individual behavior in disasters. One measure of "good theorizing" is the range and variety of phenomena that are addressed (Schrag 1967; Dubin 1978: 138–141).

Still another important aspect of Stallings' work is his placement of disasters among other contexts in which human behavior might be studied. Dynes also alluded to this issue. Stallings advocates approaching this problem using the mechanism of taxonomy, and in doing so clarifies one part of the theoretical challenge facing disaster researchers. At a general level, he points out that disasters are one kind of exception in a class characterized by uncertainty, indicating an awareness that many sets of conditions can represent stresses without meeting the traditional meaning of disasters. Developing typologies to identify and isolate these different kinds of settings facilitates theoretical development and offers solutions, for example to the challenge of what to do with conflicted situations. Stallings (1988) has previously examined the issue of whether such situations ought to be studied under the rubric of disaster. Using taxonomic reasoning one can create theoretical logic (manifest in a typology) that acknowledges the overlap in some characteristics of conflicted situations with disasters, but places them in a different category of collective stress (or uncertainty situations). Quarantelli (1986) similarly has suggested the need for a classification of crisis situations, of which disasters might be one type.

Stallings also indicates a role for other typologies dealing with the specific category of disasters. He points out that the physical character of disasters plays not necessarily a role in theory development, but does play a role in, among other things, site selection by researchers and policy development by practitioners. Some physical scientists also use physical characteristics as a means of setting parameters for their research, and many citizens and non-scientists think of disasters in phenotypic categories. Stallings' point is that the physical side of disasters will not and should not go away, and one way of systematizing the issue is to develop typologies. A range of typologies already exist that extend from very descriptive (unidimensional and thus technically not a typology) to more complex classifications based on different aspects of agent characteristics (Perry 1989b). Certainly, one would not be confined to developing typologies of disaster settings either. What has been called *generic functions* in emergency response—warning, evacuation, emergency operations centers, sheltering, and so on—are natural targets for typology construction on which some work has already appeared in the literature.

In addition to his definition of disaster, Anthony Oliver-Smith proposes at least two ideas on which I would like to offer comment. His focus for discussion is upon societies—not unusual considering his prominence as an anthropologist. He elaborates his definition of disaster by noting that it may be seen as "symptomatic of the condition of a society's total adaptational strategy." Thus, his elaborations address questions of the origins and consequences of disasters.

Oliver-Smith develops an ecological perspective on disasters that reminds us that society and the environment not only overlap periodically (sometimes to produce disaster), but interact in a fundamental way to "produce" one another. Many have pointed out that "disaster" can only occur when the human use system intersects with nature: if there had been no people around, Mount St Helens would have been "just an eruption." Oliver-Smith challenges us to think theoretically about the idea that human societies both are shaped by the environment and shape it. Clearly, this forces one to think more broadly about the origin and correctives associated with disaster. When he speaks of "socially produced vulnerability" the case that comes quickly to mind is that of hazardous materials. In generating chemicals, we modify the environment. In the way we package, transport, store, and utilize the chemicals, we produce vulnerability. When the social occasion of a disaster takes place involving chemicals, we have an opportunity to adapt. If an attempt is made at adaptation, it might be pursued by generating changes in the environment. The changes could be fundamental (as in the decision to stop creating the chemical) or they could be related to the state of technology and knowledge regarding the chemical (creating breach-proof packaging, substituting another less dangerous chemical in the same process, or devising protection for those exposed).

The point is that humans and the environment do not just coexist but they remake each other (to risk anthropomorphizing). This has important theoretical implications for how we go about understanding (and studying) images of cause and effect. And these causal images affect the possibility and efficacy of what are traditionally called preparedness and mitigation measures. Decades ago, a colleague from the University of Colorado told me that "water has a gravity-like attraction to mobile homes; rivers migrate toward them, so they should all be manufactured with pontoons." Indeed, this tongue-in-cheek vision of cause and effect does imply an interesting manufacturing solution, while ruling out certain types of social management solutions (e.g., if water chases trailers, zoning restrictions are no help). With mitigation occupying a major place in the agenda of the Federal Emergency Management Agency as well as among disaster researchers, the issue of origins becomes a critical point of departure for research and policy. More effective and grounded conceptions of origins will lead to more efficacious research and policy.

Oliver-Smith also pursues the notion that vulnerability and consequences of disasters are potentially unevenly distributed across social groups. Indeed, the claim or observation has been made before. What Oliver-Smith offers is a systemic means of understanding why the unevenness exists and perhaps how one could correct it. This is approached through the notions of political economy: a call to understand the interactions of social, political, and economic forces, and their impact on what is defined as disaster, what mitigations are available to whom, and what consequences flow to whom. In the US recently there have been charges of "environmental racism", and certainly students of disasters in other countries and cultures have reported over the years on differential distribution of outcomes relative to impact consequences and receipt of aid. Oliver-Smith's notion subsumes these issues and encourages the development of a broader framework for understanding and interpreting all aspects of disasters. Quarantelli (1995b) has pointed out that collectively disaster researchers have not systematically accounted for the larger social context in research design and theory. Certainly, careful attention to political economy as suggested by Oliver-Smith represents an appropriate corrective strategy. Finally, the development and systematization of a political economy perspective offers an important tool to those who conduct cross-national (cross-cultural) research and wish to begin to build (and eventually test) models and theories. Namely, a political economy approach can serve as the basis for creating classification schemes that allow tests and comparisons across cultures and nations that are grouped (to eliminate error variance not associated with theoretical contentions) in meaningful ways.

Uriel Rosenthal's paper, in my mind, raises at least two critical points with far-reaching implications. First, he adopts a process orientation that traces both causes and effects through time and through the social structure. He uses social structure broadly, because his notions are meant to be global rather than just societal or regional. There are several very provocative implications of the way Rosenthal presents and discusses the process orientation. One of the most striking for me was the "reconceptualization" of the notion of sudden onset disasters. If one is analytically able to carry the process orientation through, sudden onset disasters become very rare. This occurs because Rosenthal emphasizes "tracing" the disaster and allows for multiple simultaneous specifications of causes and consequences. A dam collapse—usually thought of as sudden onset—becomes less so because Rosenthal would want to examine the quality of construction, the politics, and personages of its erection, decisions to locate or not locate people in the path, and the kinds of channelization for drainage, among many other factors. There are both structural and non-structural aspects mentioned here, each with different places in social time, and virtually all taking place long before the dam fell.

I believe Rosenthal's thinking is much deeper than just making a point about sudden versus gradual onset notions, which highlights a second implication of this process notion. His use of the idea of "creeping disasters" begins to capture his belief that one must think about disasters with complex and interrelated origins as well as consequences. In part, he sees disasters as without clear beginning or end; to find a beginning, you trace back as far as you can through as many "channels" as possible, but the "reality" of finding an "origin" (or an "end" for that matter) is at best theoretical. Of course, tracing multiple origins or multiple consequences is not completely new in disaster research. In a narrower sense, we have known about "secondary impacts" for decades; for example, the idea that earthquakes "cause" or are associated with fires, hazardous materials, accidents, and so forth (cf. Lindell and Perry 1996). These contentions usually aim only at contemporary consequences and are narrowly seen in geography and time. Rosenthal is much broader, considering causes and consequences, across time and across social arrangements. The only recent treatment of disaster issues that appears to approach the use of process in the way Rosenthal intends it is the Report on Mitigation recently completed in connection with the Second Assessment of Research on Natural Hazards through the University of Colorado at Boulder (Lindell 1997).

At one level, it is difficult to see exactly how Rosenthal's broader vision would be translated directly into research; the designs would become too cumbersome to execute. At still another level, however, there is much to recommend his message: disaster research would benefit immensely from broadening the way we conceive of both origins and consequences, and existing narrow designs can be effectively expanded. A second important theme in Rosenthal's work is an explicit concern with future disasters. He urges us not to just think of what agents might be involved, but to identify definitional *dimensions* of such occasions. In doing this, Rosenthal alerts disaster researchers to what may become new and necessary research design features, as well as cueing us to begin to expand our theoretical thinking to accommodate the dimensions. He elaborates three dimensions, induced from his own research and that of others. Rosenthal argues that future disasters will be *transnational* (global); not just in impact but also in their origins. He also emphasizes the *mediazation* of future disasters, emphasizing the many roles that mass media will play. He points out that media impacts will be (and have been) active in defining disasters, victims, origins, and impacts; with the notion that media at some point may be able to create disasters or may participate in ignoring them. Finally, he argues that *politization* will be an ever-increasing aspect of future disasters.

Steve Kroll-Smith/Gunter raise a variety of epistemological questions in their paper. One issue related to the definition of disaster merits comment here. They underscore the importance of listening to "voices of people and communities who live through, with and among disasters," presumably as a means of learning about the definitional content of the concept on an experiential level. I infer that Kroll-Smith/Gunter are talking about the importance of induction, which I place within the tradition of sociology as a social science (Berger, Wagner, and Zelditch 1989). Even if Kroll-Smith/Gunter are advocating an (unspecified) alternate epistemology, I think the comment is important for and has meaning within a structure of the postulates of science (Sjoberg and Nett 1967).

In seeking to understand a phenomenon, one can begin with simple observation in the world of experience. As one (or more) observers accumulate observations, we seek to identify patterns even though this may require much time and many observations. If patterns become detectable and identifiable, one groups them—hopefully extracting elements of meaning—and begins to generalize and gather these elements into concepts. This conceptualization is a primarily inductive process, generally captured under the rubric of the "logic of discovery" that forms one basis for scientific description (Blalock 1969). When one elaborates a concept, by necessity, we are creating a (nominal) definition (Papineau 1978: 21).

Continued observations serve as a means of refining and revising a concept. Also, at some point the observer moves on to create propositions by linking concepts and to create theoretical edifices by linking propositions. The objective in these endeavors is explanation, not just description, of the phenomena that we observe. Once we get to the theory stage (an activity of assembling both our knowledge and our speculations), we begin a deductive path ("the logic of proof") of creating hypotheses for testing in the world of experience to fulfill the empirical requirement of the scientific method. The process never stops, however, and at least I see change as inherent over time (whether changes occur via something like Kuhn's revolutions or via an accumulation of knowledge phenomenon, is fortunately not an issue here).

To return to the point by Kroll-Smith/Gunter, one should not look at conceptualization as a unidirectional, finite process that just ends as one moves towards concern with explanation. There is a backlog of many case studies, representing many observations over many years, that focus on many aspects (dimensions) of disaster experience using many different units of analysis. Based upon this experience, many researchers are comfortable with definitions and find them necessary for the theory building process to begin (as Kroll-Smith/Gunter seem to indicate). Other observers can certainly choose to continue to focus on more inductive strategies with many different aims, including enhancing conceptualizations. In all, there should be mutual respect for each activity. Moving on to developing explanations should not exclude reassessments of theoretical thinking and concepts: looking for new dimensions, new uses, new or novel interpretations (cf. Glaser and Strauss 1967). Because social science embraces an empirical assumption, the test of the utility of explanations (derived from theories and linked to our descriptive ability) lies in the extent to which they match (or predict correctly) behavior in the world of experience.

The encouragement and respect of knowledge generated in many different ways is important for the growth of sociology. While differences in fundamental postulates (world-views) make it difficult for genuinely different knowledge systems (epistemologies) to share information, certainly conclusions can be shared (cf. Hindess 1977: 211). Within the philosophy of science, metatheoretical process virtually demands that inductive and deductive strategies not just respect but complement and cultivate each other (Perry 1989b). Randall Collins states the issue very concisely:

> Much of what we express today about each other's work is negativistic, hostile, dismissive. This factionalism is debilitating because we need multiple approaches in order to cross-validate our findings . . . For sociology to make progress, we need some spirit of generosity, instead of a spirit of factional antagonism.
>
> (1989: 137)

CONCLUSIONS

The discussion of the definitions and contexts of disaster by Dynes, Stallings, Oliver-Smith, Rosenthal, and Kroll-Smith/Gunter reflect different perspectives and have implications for the future of disaster research and the accumulation of knowledge regarding disaster behavior. In closing, I shall briefly reiterate and interrelate selected points made by the authors and include some of my own observations.

I saw much consistency among the definitions offered by Dynes, Stallings, Oliver-Smith, and Rosenthal, as well as some overlap with the definitions offered by the first authors in this volume, and selected definitions historically used in the field. To achieve this vision of relative consistency, I followed Quarantelli's (1982) well-founded warnings to separate aspects of definitional statements that dealt with antecedent conditions, subsequent consequences, and phenotypic characteristics of disasters. All of these features—from the way they are identified, through their descriptive elements, to differential attributions of relative importance—rest somewhat upon researchers' idiosyncratic experience and emphases, but more importantly upon theoretical rationales. In a sense, this kind of information tells us *about* disasters, rather than answering the query "What is one?". One should expect theoretical rationales and assumptions especially to differ (in small or large ways), if only as a function of the way social science operates (cf. von Bretzel and Nagasawa 1977). Hence, when components addressing origins, consequences, and characteristics are included in definitions, and in fact are different, one should not necessarily conclude that the definitions themselves are different.

The matter of reacting to variation in definitions of disaster was mentioned by two of the authors and certainly is a question worth addressing. Both Oliver-Smith and Kroll-Smith/Gunter pointed out that controversy has raged for decades in other disciplines regarding the definition of key concepts, with apparently minor consequences for knowledge acquisition and theoretical growth. Oliver-Smith makes an argument that disaster may just be a contested concept. Blalock (1984: 62) has indicated that controversy over concepts and definitions sometimes occurs when social scientists try to precisely capture what is ultimately a "fuzzy reality."

My reaction is to agree that disaster is a multidimensional concept. In some cases one can expect that by their very nature definitions will differ; definers (ideally) make theoretical assumptions and tend to pack additional elaborations of origins, characteristics, and consequences into their statements. Variation stemming from these reasons is probably minimally problematic. It certainly indicates that abstract (theoretical) thinking is going on. In evaluating definitions, however, it is reasonable to strip away elaborations and concentrate on what scholars say "is" the target phenomenon. To the extent that definitions do vary on fundamentals, we should look carefully at the intended referents. Are they isolating different phenomena-to-be-explained, or specifying more dimensions of the same multidimensional concept? In one case it may be necessary to revise our definition, while in the former we may need to add additional definitions, create new typologies or reformulate existing typologies. The presence of multiple different definitions would seem to be something that would be handled (or possibly ignored) in the context of ongoing research and theory.

In reading the author's definitions, I found reasonable consensus that disaster is a socially defined occasion, serving as a context for human behavior, recognized across social time as a radical change in the effectiveness of social structures (norms, practices, beliefs, etc.) to meet human needs, and framed in a social change perspective. To arrive at this notion, certainly I have inferred and interpreted and looked at each author's full statement instead of just the "definition."

The mention of social change perspectives does require some qualification. All of the authors—some more explicitly, some less—spoke of disaster in reference to *change*. Stallings builds his framework around the notion of a change in the social structure (amalgamation of routines) that in turn invites (not necessarily demands) changes in human, organizational, or institutional behavior. In passing, Stallings acknowledges that some may see this as an "equilibrium" theory in the tradition of very old systems theory thinking. It appears to me that the theoretical nature of routines, exceptions, and exception routines clearly does not fit the (dated and simplistic) notion of a system which is "restored" to its "original" state following an exception. Dynes even more explicitly places disaster in the realm of social change. By framing disaster as an index of a society's adaptational ability, Oliver-Smith also embraces a social change perspective. By emphasizing *process* in the definition, as well as the specification of causes, characteristics, and consequences, Rosenthal makes social change a central feature of his thinking.

In the literature, there are contrasts of social constructionist views with conceptions of disasters as non-routine social problems founded in functional thinking (Drabek 1989a; Stallings 1991, 1995; Kreps and Drabek 1996). Implicitly, one might sense intellectual disagreement regarding the capture of disaster research in a social change perspective versus a social problems framework. I do not see the two ideas as mutually exclusive. Certainly social constructionist thinking is distinct as a theoretical orientation from the functionalism cited in discussions of disasters as non-routine social problems. On the other hand, Oliver-Smith—I think successfully—*defines* disaster in terms of adaptation and change, and then *elaborates* particular disasters in view of socially manufactured vulnerability (an unmistakably social problems notion). Rosenthal makes a very similar point when he not only discusses disasters whose origins could be socially manufactured, but also reviews the notion that consequences can also be socially manufactured. It appears to me that social change tends to be treated as an inherent feature for definitional purposes. The elegant statement of disasters as non-routine social problems by Kreps and Drabek (1996) emphasizes "non-routine" which begs comparison to "routine" and can be understood in a broader rubric of change from one state to the other. When we extend beyond the definitional issue, however, whether one embraces social constructionist or functional "theory" as a basis for understanding the causes, consequences, or characteristics of disaster is part of a different problem: explanation. The choice between the perspectives would seem to be appropriately based on theoretical preference and analysis or assessment of the empirical record. Neither choice precludes acknowledgment of the notion of change in defining disaster. Of course, a critic could reasonably point out that I am treating social change in a very abstract fashion. At the same time, I am allowing much leeway in the claims that there really is an edifice one might recognize and label "formal theory" connected with functionalism or social constructionism.

In both my discussion and the author's papers, there has been a tendency to seek a general or abstract definition of disaster. There is an absolutely critical issue that can be found in the implications of this abstractness. It was mentioned by both Dynes and Stallings, and I think it needs to be emphasized and underscored. For me, this issue is: Are we defining "disasters" or some more abstract class of phenomena (Stallings called it uncertainty, Barton used collective stress situations, Quarantelli has suggested crises) *of which disasters are just one type?*

I see great value in a general or generic (nominal) definition that potentially includes a variety of phenomena. Possessing such, one can proceed to more precisely delineate the phenomena of interest through taxonomy. This process has the advantage of being founded in theoretical thinking that can yield many typologies along both phenotypic (surface or visible characteristics of the phenomenon) and genotypic (more fundamental "theory" bases) lines. The question is where—intellectually and theoretically—do the definitions proposed here fit? Do they reflect a vision of "collective stress situations" or "disasters?" Dynes and Stallings seem to have (self-identified) their definitions as part of the more general category without explicitly describing the other "types" in that classification. Rosenthal also appears to be addressing many of what would be called "types" of disasters—at one point separates out "accidents" himself—and thus would appear to fall into a collective stress situation category. Of course, one cannot solve all the problems of the field in a single paper. Expressing awareness of the larger category of "stress" phenomena is a theoretical gain in itself; constructive definition of disaster can be (and was) undertaken without exactly specifying the higher level of abstraction. Indeed, Dynes achieves his most precise specification of disasters through a typology which itself might be one or more cells of the more abstract typology.

Ultimately, however, possession of a clearer picture of the more abstract or general case will allow us to focus more precisely as we create definitions of disaster and as we decide what concepts related to those definitions of disaster are appropriate targets for taxonomic thinking and the generation of typologies. To untangle the intellectual and research issues involved, disaster researchers need not just define and specify types of disasters, but more clearly place disasters into the broader class of phenomena as well. Indeed, explication of the more general class can serve as a way of acknowledging and theoretically separating from disasters phenomena that sometimes get included but do not actually fit on various grounds (conflicted situations or very diffuse phenomena, for example).

The preceding discussion raises another definitional matter for disaster research. Quarantelli (1986) has pointed out that deciding what disasters "are" is not just a matter of fitting a conceptual envelope around phenomena that have been labeled (by someone) disasters in the past. While we do need to clarify matters, such as the place of conflicted situations, we also need to deal with (classify) contemporary phenomena such as the ecological threats mentioned here by Oliver-Smith and the threats that arise in connection with rapidly growing computer technology mentioned by Quarantelli in the introduction to this volume. Clearly, the ability to identify which of many potential phenomena-to-be-explained are appropriate targets for attention is critical for the growth of any theoretical superstructure.

Certainly the author's definitions and discussions reflect awareness of Quarantelli's concern. Rosenthal addresses it most explicitly, devoting much of his elaboration to such events and even laying out dimensions of future disasters. The typology created by Dynes also takes a direct approach in that there are "types" that subsume a wide range of occasions including conflict or violent situations, epidemics, and technology-related events (sector/network disasters); clearly many kinds of phenomena would be captured. Also, the logics and frameworks developed by Stallings and Oliver-Smith are fertile ground for taxonomic thinking which could produce classifications that capture existing disasters and anticipate the future. Ultimately—acknowledging again that it would be inappropriate to expect the authors to solve all dilemmas in a single stroke—I believe that the discussions by these authors provide two important advances in the direction of Quarantelli's concern. Dynes' typology, as it stands, provides one means of classification that encompasses the things Quarantelli cites. Of similar importance, Stallings and Oliver-Smith provide *foundations* for accomplishing the same feat.

Still another issue that drew the attention of the authors was the question of who defines disasters and what comes of it. Everyone can define disasters. My comments here have been made in the context of social scientists defining disasters as a means of placing intellectual parameters on their field of study. These kinds of definitions have implications for what knowledge is sought, how it is collected, how it is stored, and how it is evaluated and integrated relative to understanding human behavior.

Many people and groups both define and need definitions of disaster, yet put the definitions to very different uses than social scientists. Stallings very clearly makes this point and proposes a research agenda to collect information on such definitions. One would expect possibly radical differences between definitions developed by some groups and those used by social scientists. Certainly social science definitions are (and should be) related to others definitions; hopefully each informs the others. But each group or individual creates a definition with different ends in mind. When social science definitions are used in description and explanation one expects that they will conform to social science standards. In the tradition of taxonomy in sociology, Stallings makes the distinction by referring to definitions devised for social scientific purposes as "constructed," labeling definitions created by other actors "existential."

Finally, it is important to appreciate the relationship of defining disaster to the future of studying disasters. Operating within the framework of social science, disaster researchers share the professional goals of description and explanation (Homans 1967). As Drabek (1986) and Quarantelli (1995b) have shown, disaster research has many examples of description, but our excursions into explanation have been relatively few. Certainly creating new models and formal theories, or extending existing social science theoretical work to capture disaster phenomena, are actions that move disaster research further in the direction of explaining and predicting. It is also important to acknowledge that taxonomic thinking itself is a theoretical activity (Turner 1989). In producing typologies to help with the social scientific specification of the meaning of disaster, we are engaging in theoretical development. Part of the utility of typologies is that, even if a researcher fails to classify empirical findings, provided enough information is reported, theorists or other researchers can subsequently place the findings within existing typologies. In that sense they are both forward looking—they can serve as a guide to what can be studied—and backward looking—they permit the organization of research findings generated in the past. Consequently, as we develop and refine definitions, and elaborate and specify them through the creation of classifications, we are not just identifying parameters for disaster research, we are expanding the theoretical capacity of the field, thereby adding to a theoretical superstructure.

REACTIONS TO REACTION

17

ON THE CONCEPT OF DISASTER

A response, a slight recovery, and less reconstruction

Russell R. Dynes

My primary response to Ronald Perry's comments is appreciative for his effort, integrating diverse visions of disaster. Two of the comments he made about my paper can be noted with agreement. He suggests that the typology of disaster that I developed based on the *community* might be better framed in terms of *social systems*. I agree, but I chose community as a more pragmatic referent for the more abstract social system. In retrospect, community is also abstract so my initial decision was not particularly helpful.

Perry also suggested that the concept "disaster" might better be nested in a larger category of "collective stress." I have always admired Alan Barton's creativity and audacity in developing, in the late 1960s, that classification system which attempted to bring disaster into a more inclusive typology reflecting types of social impact. Barton argued that collective stress occurs "when members of a social system fail to receive expected conditions of life from the system" (1969: 38). That has a nice ring to it, until we try to imagine what that might mean and how we might develop indicators for stress. What are "expected conditions of life?"

Posed in that way, there are implicit notions of how social systems are supposed to perform (reminiscent of functional prerequisites within Parsonian structure functionalism, popular at the time when Barton wrote) and there are echoes of some base line of psychological well-being. Are people supposed to have predictability and stability in life, the lack of pain and injury, the absence of fear and anxiety, and the assurance of the fulfillment of aspirations and expectations? If so, disaster mitigation would only require reduced expectations. Defined in those terms, collective stress is a constant condition of routine social life.

Barton rather creatively took three dimensions of disaster agents—scope of impact, speed on onset, and duration of impact—as *causes* of collective stress and suggested that such stress involved unfavorable changes in the external environment as well as internal–social disorganization. Again, this has considerable face validity, but it does not directly address indicators of collective stress. Unless we make progress in defining that, we will continue to use imperfect surrogates. The fact that the research field has not moved very far in that direction points to our own lack of sociological imagination, as well as the persistent difficulty in developing measures that are not individually and psychologically based. This is why I tried to use organizational involvement as the primary base for the development of my typology. Indications of "extraordinary effort" attempts to point to behavioral indicators which could differentiate disaster behavior from "normal, routine" behavior.

After framing the original paper, there was still time to recover and to consider my initial effort and its shortcomings. Some of those afterthoughts might be relevant here. In some instances, conceptual analysis can be illuminating, but it also can be paralyzing. Conventional wisdom emphasizes that adequate theory should precede research, but never what should precede theory. Such preoccupations often prevent us from raising realistic empirical questions. One of my colleagues spent his entire career trying to conceptualize a particular segment of his field. When he finished, his solution was passé and so was he.

Conceptual analysis is especially difficult when dealing with popular and commonplace terms. Attempts to gain precision seem pretentious to others. In addition, it is not necessarily true that conceptual consensus will lead to an explosion of productive research. More likely to spur disaster research would be an increase in research funding, produced by media and political attention.

Taxonomic approaches always produce residual categories. The further elaboration of taxonomies to clarify those "contaminated categories" usually leads to cells, empty of content. So, increased clarity can lead to decreased significance.

Many putative theoretical breakthroughs in disaster research are based on what I would call *agent exceptionalism*. In these instances, researchers suggest that "their" disaster is different—bigger, more important, more significant, more traumatic—than those others have studied. Such claims are usually true by assertion. If you define disaster by qualities of agents, different agents produce different disasters. While we need to appreciate that some writing and research careers are built on making tautological differentiations, in the long run, such claims add little to our collective understanding.

It is not likely that we will make significant progress until we have developed some sociological measures of social impact/social stress. My suggestion of "extraordinary effort on the part of community organizations" might be a start, since it gets away from agent characteristics. I would infer from Perry's comments that Stallings suggestion of "disruption of routines" points to a similar direction. Unless we develop independent measures of "social" impact, the concept will remain the functional equivalent of post-traumatic stress syndrome, caused by everything and having consequences for everything, therefore, explaining nothing.

Having set the directions that others should follow allows me to be optimistic for future progress anticipating some younger scholar will solve those conceptual problems. That optimism was shaken recently when the fortune cookie presented to me with my bill at the nearby Vietnamese restaurant, proclaimed that "In youth and beauty, wisdom is rare." While that tempered my optimism, it also provided me with an explanation for my own failures.

18

REPLY TO PERRY'S REACTION PAPER

Robert A. Stallings

Perry's perceptive and generous comments on my chapter, "Disaster and the Theory of Social Order," give me little with which to disagree. Therefore, I will expand briefly on five themes that appear in his reaction paper: development of a theoretical superstructure; the role of taxonomy in such a theoretical project; differentiation between type and subtypes; levels of analysis; and the relationship between social change and exceptions to routines.

First, it is the second element in the title of Perry's reaction paper that is central to this discussion, in my view. Not "Definitions of Disaster", but "The Development of a Theoretical Superstructure for Disaster Research" is the more important of the two elements. I do not deny the relevance of definition for guiding what we study. Nor do I deny the importance of definition for being able to generalize from our findings. However, it is easier to resolve problems of definition and generalizability from the standpoint of a general theory than it is to resolve them in isolation, without reference to such a theory.

Second, Perry invites me and some of the other contributors to this volume to extend our efforts through "the mechanism of taxonomy." I confess to being less persuaded of the benefits of taxonomic development than is Kreps (1989c), for example (see also Turner 1989). Too often taxonomy becomes an end in itself. I would prefer that we simply go about the business of our empirical work, letting typologies emerge inductively as we need them in order to make comparisons among the things we have already studied or to identify the things we should study. Among the distinctions that I personally think would be theoretically profitable to pursue are these two: exceptions to routines that are identified by insiders (i.e., participants in those routines) versus exceptions that are identified, initially at least, by outsiders, especially by experts; and exceptions to routines that are defined as existing in the present versus those claimed to be arriving in the future (i.e., things which "threaten future generations"). I see no need to derive in advance a typology from these distinctions, such as by cross-classifying them.

For my purposes, the conventional sociological delineation of the major institutional sectors of society provides a sufficient taxonomic starting point for future inquiries. Why not identify and compare what constitute exceptions to routines across different institutions? What are the things that are said to threaten the status quo in a fundamental way in each sphere of life? In other words, what are the similarities and differences among threats to the political, economic, religious, scientific, educational, and familial spheres in a given society? What are the similarities and differences in the way defenders of valued routines in these spheres try to manage such disruptions? What do they tell us about how societies deal with disasters?

As one example, some Swiss banks and the government of Switzerland have been criticized recently for certain financial transactions with the government of Germany between 1933 and 1945. Evidently, the nature of their international reputation is important to the success of Swiss banks, and to the financial health of the Swiss government. Damage to this reputation is a potential "disaster" for the economy of Switzerland and the banking practices (routines) upon which it rests. An examination of this controversy and a comparison of it with responses to "real" disasters should not only illuminate the way societies deal with the latter, but also contribute to a more general theory. The conventional "taxonomic" distinctions among institutions seem perfectly capable of facilitating such a project. More importantly, this Swiss banking example suggests the importance of a different kind of comparative research than that so frequently, and correctly, urged upon disaster researchers (e.g., Taylor 1978). Not only the comparison of disasters across nation-states, but also the comparison of substantively different types of events, each having in common the fact that they interfere in a serious way with the highly valued routines comprising a major sphere of social life, would be beneficial.

Third, there is the issue of the differentiation between type and subtypes. I do not think of this as a matter of taxonomy, although others may. In my earlier chapter, I argued for a theoretical project that broadened the study of disasters to an inquiry into the fundamental processes of social order. To facilitate this without creating a priori an entirely new theoretical superstructure, I proposed a focus on routinization as the continual effort to maintain order in a collective existence characterized by change, discontinuity, and difference. In search of a generic term, one that would avoid the implications of terms such as "breakdown," "disruption," or even "interruption," I hit upon the term exception. Disasters are one of its subtypes. Perry's reaction paper stimulated me to think further about relationships between the term exception and some others that might also be used to subsume the term disaster.

Quarantelli's distinction between "catastrophes" and "disasters" (1995b: 8) is an extremely useful one, but for a different purpose. Both terms refer to substantively similar phenomena which differ in scale in some way, such as in scope of destructiveness. This difference in scale encompasses qualitative as well as quantitative dimensions of the consequences of events, but not fundamental differences in their substance. Hence, I do not consider catastrophes and exceptions to be synonyms; both catastrophes and disasters are subtypes of exceptions. Likewise, Luhmann's (1993) use of the term "uncertainty" is related to what I am calling exceptions, but it too is not a synonym. Exceptions are threats to routines; uncertainty is an outcome or consequence of exceptions which is to be avoided through improvisation or implementation of exception routines. Uncertainty is the ultimate potential threat posed by an exception, not the exception itself. Closer in meaning to exception is the term "crisis." Although as loosely used colloquially as the term "disaster," it conveys the same sense of being at a crossroads that an exception represents. Routines are threatened, and the threat cannot be ignored nor action postponed; there is the belief that things will never be the same unless the exception is addressed. For example, the political and economic issues surrounding past financial transactions between Switzerland and National-Socialist Germany will not go away; something must be done to resolve the "crisis." Exceptions are crises for social units, not in themselves but because certain people see them as such.

Fourth, as Perry suggests, examination of exceptions and exception routines can take place at any level of analysis. "Disaster," it seems to me, has specific legal, journalistic, and public referents that makes it only a colloquialism to call the divorce of a single couple a "disaster" or even to call divorce rates at a given time and place a "disaster." While the theory I favor is one at the macro level, in the sense of being about societies, its propositions apply to any of the subunits of societies.

Fifth, and finally, the relationship between change and routinization needs elaboration in order to further develop a theory of social order. Here, I offer only a few thoughts about one aspect of this relationship, that between change and exception. I offered the proposition that a routine invites its own unique exceptions. For example, modes of transportation using internal combustion engines are vulnerable to disruption in ways that modes of transportation relying on animate sources of locomotion are not. A sudden decline in the availability of petroleum-based fuels disrupts an economy based on automobile, truck, and airline transportation, but has little impact on one for which animals provide the chief means of transportation. This proposition is not just about technological breakdown; for instance, the nuclear family is vulnerable to types of breakdown that the extended family is not (and vice versa).

The point is that the relationship between change and exception is a complex one. Neither "optimistic" nor "pessimistic" theories of social change provide satisfactory explanation. Optimistic theories posit crises as opportunities for improvement in addition to being times of destruction and loss. Exponents include Sorokin (1942) and Carr (1932); both treat crises as analogous to the way Weber (1978: 1111–1157) viewed charismatic leaders as antidotes for increasing rationalization. Pessimistic theories, on the other hand, posit crises as the negative consequences of social change. Exponents include Marx (e.g., 1964: 231–258) and Beck (1992). Especially for Beck, the future looks bleak because of changes in the past. Neither optimistic nor pessimistic theories, however, capture what I see as the relationship between change and exception. Clausen (1983) comes closest to offering a useful approach to this complex relationship with his FAKKEL model.

Every routine (i.e., structure) is vulnerable to disruption. The range of such disruptions is limited and is determined by the nature of the routine itself. The most important routines are "buffered" (to borrow Thompson's 1967: 19–24 usage) as best as can be managed. However, new routines are vulnerable to disruption in ways that previous ones were not. For instance, monotheism constituted a challenge to polytheism in a way that was totally different from the challenge that Martin Luther presented to the church in Rome. Exception routines that would have worked in the one case would not serve well in the other. Over time, exception routines that work become a part of the routines that they buffer. Disasters, like religious revolutions, represent excellent opportunities not merely to study reactions to stress but, more fundamentally, to understand the processes through which societies attempt to maintain order in the face of what looks like chaos.

19

COMMENTS ON PERRY'S COMMENTS

Uriel Rosenthal

TELLING ABOUT DISASTERS

In reading Perry's reaction paper, I feel good to have shared in an effort to generate information that "tells us *about* disasters, rather than answering the query: What is one?" This fits fully into my notion of success in an international exercise to conceptualize disasters. Perry pays considerable attention to the nominalistic elements of conceptualization, but he withstands the temptation to construct a strictly fabricated consolidated definition of disaster. One could say that this testifies to the gradual development of a mature body of knowledge that by now does not need to prove itself through extensive jargon and technical vocabulary.

CHANGE

My way of looking at disasters corresponds with Perry's idea in his reaction chapter that:

> disaster is a socially defined occasion, serving as a context for human behavior, recognized across social time as a radical change in the effectiveness of social structures . . . and framed in a social change perspective.

Two points should be raised. First, Perry borrows from Dynes the notion of disaster as an occasion for nonroutine action. It is interesting that this perspective recurs to what sociologists once used to call the latent function of disaster. Disasters will mobilize the collective energy and will activate dormant norms and values. They may do away with daily routines and bureaucracy. In that sense, disasters bring into the open the forces hidden within the social fabric. It usually comes as a surprise when communities turn out to be able to cope with the initial impact of disastrous events without resorting to outside assistance. Ostensibly, communities may be much stronger and action-oriented than many authorities and observers would assume them to be.

Second, as Perry in his reaction chapter says, the perspective of social change is definitely crucial to my line of thought. Indeed, it would be difficult for me to have common ground with scholars who would not accept disaster as process. I understand Perry's concern about the complexity of a full-fledged perspective of change and process which embraces both causes (conditions) and short- as well as long-term effects. It may become even more complex when the interrelations, feedback and feedforward mechanisms, and circular loops of contemporary and future disasters are taken into account. But let us face it: unfortunately, simple concepts and easy solutions will not do in a world of complex disasters. For that reason, we should at least upgrade Dynes's concept of *compound disasters*, be fully aware of Erikson's (1976) "disaster after the disaster," and pay ample attention to the ever-increasing subjectivity of the disaster label. When CNN declares it to be a disaster, it is a disaster in its consequences.

A GENERIC DEFINITION: HOW ABOUT CRISIS?

Once more, I enjoy being on board with the other colleagues in what Perry in his reaction chapter calls "an absolutely critical issue." Quoting Perry "are we defining 'disasters' or some more abstract class of phenomena . . . *of which disasters are just one type*". It is just pleasant not only to be on board, but to be side by side with Quarantelli who has suggested in the last chapter in this volume that we explore crisis as the generic concept. I could not agree more. It is indeed impossible to study the above-mentioned complexities of contemporary and future disasters without taking a more comprehensive stance. As our *Crisis Research Center* at Leiden in the Netherlands has experienced over the last fifteen years, the concept of crisis, which links threat, uncertainty, urgency, and stress, is particularly useful as the common denominator for a wide variety of phenomena, one of which is disaster (Rosenthal, Kouzmin, and 't Hart 1989). It cannot be coincidental that the various generic notions developed by other scholars can easily be molded into the conceptual framework we tend to apply. For example, Stalling's emphasis on uncertainty is part of the generally accepted definition of crisis. Barton's notion of collective stress (1969) is very apt in linking crisis and disaster. Non-routine response, which is among the more important ingredients of disaster, is an almost logical derivative of critical uncertainty and tends to emerge as one of the dominant propositions of crisis decision-making.

As the comprehensive agenda of the *Journal of Contingencies and Crisis Management* suggests, opting for crisis as a generic concept creates new opportunities for the exchange of knowledge between hitherto separated scholarly worlds (Rosenthal and Kouzmin 1993). Although once popular among the founders of the social sciences, for a considerable period of time, crisis used to be the pet concept of experts in international relations. They kept the concept for themselves, demarcating the empirical domain by exclusive reference to the threat of war, in particular nuclear war (Lebow 1981). They did not appreciate the potentialities of a wider application of the concept and crisis-derived hypotheses and propositions, leaving such efforts to others. On the other hand, those working in other domains (disasters, riots and turmoil, terrorism and highjackings, ecological threat, corporate disturbances such as produce recall) developed their own bodies of knowledge or, as in the case of disaster research, had already settled for their own terms, concepts, and theoretical perspectives.

With the end of the Cold War, clear-cut nuclear threat has given way to an overall sense of uncertainty. The context of international crisis management has changed dramatically. Nowadays it encompasses a mixture of international peace-keeping and peace-enforcing operations, on the one hand, and large-scale relief to disaster-prone or disaster-stricken areas on the other hand. Future disasters will be part of the world of crises. They will be transnational, mediatized, and highly politicized. For that matter, future disasters will be genuine crises.

20

A PLEA FOR HETERODOXY

Response to Perry's remarks

Steve Kroll-Smith and Valerie J. Gunter

If we are reading what Professor Perry intended to say regarding our paper, it appears he thinks we spoke around but never fully on the issue at hand, the definition of disaster. We will leave it to readers to decide whether or not we addressed the question of definition. A debate on this point is unnecessary as our text is in this volume. Readers, being readers, will reach their own conclusions. We would rather use our limited space to reiterate a point made in our chapter that strikes us as important to the sociology of disaster.

There is a remarkable diversity of literature, concepts and perspectives on disaster that is rarely if ever considered within the legislative tradition of disaster research. When scholars use a single definitional strategy to examine a problem or issue, limit their literature citations to one another, and do not include the work of other scholars who are defining the problem in other ways, an orthodoxy is likely to emerge. Now a doxic orientation is not necessarily bad. Alignment with a particular standpoint is a necessary part of intellectual (or political) work. But when alignment approaches obligation and orientation shades into duty it might preclude the possibility of seeing other possibilities. Perhaps at that point we will seek a final resolution in a master definition, an ultimately frustrating quest.

We would like to see a bridge between the legislative or classic tradition of disaster research and other approaches to disaster that haunt its margins in ever growing numbers. Concepts like "corrosive communities" that emerge in certain types of disaster, but not others (Freudenburg and Jones 1991), invite new research into disasters and conflicts. Emergent problems of "environmental stigma" that beset both individuals and communities (Edelstein 1991), suggest the possibilities of a social psychology of emotions and disaster. The ideas that accidents are now a normal outcome of transformative technologies (Perrow 1984), and the similar concept "manufactured risk," (Giddens 1990) highlight the intricate links between economic production and the political production of risk and disaster. The unsettling idea that the people sociologists study are likely to be studying them, makes sociological research and sociologists variables in investigations of disasters (see our article in this volume). These are a few of the concepts and theoretical orientations that invite conversation with the good research of the classical school.

We are confident there will be no single definition or definitions of disaster. How could a fragmented discipline produce a common, uncontested concept? Gender, community, power, disaster, and other core concepts will continue to orient us to key processes and problems in social life while eluding our best efforts to capture them in formal definitions. Recall Collins' good advice, "A strategic problem in theory building is to find the right guiding imagery", one that "leads us to the key processes rather than one that obscures them" (1975: 51). We close with an attempt at such imagery, though we suspect it will not satisfy the classicists. Disaster is a sensitizing concept, it is at once a research topic, an administrative decree, an ordinary language description or metaphor for human experience, and a popular genre for B movies, among other possible permutations. Why not acknowledge this truism and build on its possibilities?

21

DISASTERS, SOCIAL CHANGE, AND ADAPTIVE SYSTEMS

Anthony Oliver-Smith

Professor Perry's thoughtful and constructive commentary has given me the opportunity to expand somewhat on a number of key issues I tried to explore in my chapter. Perry, citing Stallings in this volume, notes that early researchers saw disasters as opportunities for exploring the implications of theory for basic social processes. As has been pointed out far too many times to mention, disasters constitute a relatively underexploited laboratory for testing theory in the social sciences.

Specifically, for my purposes disasters provide a highly focused lens through which to examine both the concept and societally specific modes of human adaptation. For example, given the emerging instability and unpredictability of environmental conditions (Holling 1994), it is no longer possible to simply assume the adaptive fitness of our own social system. Quite the contrary, in fact. A number of factors, including population growth and concentration, the toxicity and volatility of some technologies, and productive forms maximizing resource exploitation, have placed that fitness in some doubt.

The essentially remedial concept of sustainability, however difficult it may be to define operationally, also obliges a more acute questioning along these lines. From the standpoint of adaptation, disasters may be seen as symptomatic not only of specific weaknesses (e.g., construction in flood prone areas) in a social system, but of the overall adaptive fitness of the society's relationship to its environment.

Anthropological research has found that traditional cultures, rather than living in a "more or less continual reign of terror," as Sjoberg (1962: 361) characterized them, generally took environmental dangers into account, developing over time mitigation and risk avoidance strategies that were reasonably effective most of the time (Torry 1979). However, specific societal adaptations are now linked more then ever into global systems, those larger social contexts, the implications of which have gone generally unexamined in disaster research. The political economic constitution of (under) development has created conditions of increasing vulnerability, undermining many traditional adaptations and imposing new risks on people (Horowitz and Salem-Murdock 1987; Zaman 1994). In effect, local vulnerabilities are not always the result of local causes; there are now far more systemic features to specific vulnerabilities.

As local communities come to grips with increased vulnerabilities, they enter into new relationships with both the environment and larger social contexts, inevitably affecting the pace of social and cultural change. Furthermore, in coping with disaster impacts, communities are forced to adjust past structures and practices to altered circumstances, if only in novel forms of resistance to disaster induced changes (Oliver-Smith 1992). Indeed, the impact of a disaster may usher in a conservative resistance to new social alternatives set in motion by the disaster and its aftermath. Therefore, any discussion of a disaster and its effects on a community must consider the issues of shorter-term social organizational changes and longer-term structural adaptations involving the future well-being of the community as well as the trauma of impact.

Issues of community and group survival and well-being under normal circumstances frequently involve competing interests and contested interpretations. Any perception of risk or disaster impact that threatens that well-being almost invariably becomes itself an even more contested issue. Furthermore, as the state acquires more functions and power in local contexts, tensions emerge between the expert systems and expert knowledge employed by the state and local experience and narrative as idioms of resistance and reform (Kroll-Smith and Floyd, forthcoming).

In such contexts, disasters become defined and interpreted by various interest groups at multiple levels of the total society, many with competing agendas, and become part of a scenario in which different, often unrelated, issues are played out. Dyer's analysis of the perception of impact and need among the fishing populations of Florida and Louisiana after Hurricane Andrew's impact offers a case in point. Although "objectively," damages among both groups were similar, in south Florida where commercial fishing is marginal to the larger tourist and agricultural economies, little disaster impact among the fisherfolk was assessed by state authorities, with consequent lack of attention, material or otherwise. Conversely, in Louisiana, where the fishing economy plays a more central role, the fishing population's post disaster situation was perceived as acute (Dyer 1995). Hurricane Andrew, its impact, and its interpretations were thus framed by the economic organization of natural resource exploitation of the two states, underscoring, as Harvey has recently noted, that interpretations of the relation to nature are simultaneously interpretation of society, often involving the use of identical metaphors (1996: 174). Clearly, perceptions of disaster in Florida and Louisiana constructed and engaged the issue from perspectives embedded in political economic organization.

If disaster can be so variously defined, Perry then asks in this reaction article, as others have, if the task is to define disasters as part of some more abstract class of phenomena, whether to be called "uncertainty," "collective stress', or "crises." From the perspective of impact and response, I would agree. Research with disaster victims, political refugees, and people who have been resettled by development projects convinces me that an array of sufficiently similar effects and responses exists to posit a larger or more abstract category that encompasses a variety of disruptive events and processes. Indeed, many of the same variables and categories that we employ in analyzing disasters, such as speed of onset, warning, preparation, impact, recovery, etc. are equally employed in understanding a wide variety of what Erikson has termed a form of social or "communal trauma" (1994).

In point of fact, I currently co-direct the Displacement and Resettlement Studies Program (DRSP) at the University of Florida, focusing on training, research and consultation that address both the specificities and the commonalities of disasters, development-induced displacement and political upheaval as forms of disruption of individual and community life. The common focus in the research agenda, as well as in the six courses offered in the program, is the human response to radical alteration of the total environment—physical, sociocultural, economic, and political—in terms of upheaval, dislocation, and destruction. Consultation focuses on prevention and mitigation of the many common miseries disaster victims share with development and political refugees.

Finally, while I would agree with Professor Perry that many of the attempts at definitions of disasters in this volume share certain features in common, substantial differences exist as well. However, I think the differences lie largely in intent and goals of researchers. The arrangement and importance given in a common group of definitional features account for significant differences, reflecting more the kinds of questions different researchers seek to explore rather than a malaise of intellectual disarray. While I am generally interested in most aspects of disasters, particularly as they affect practice, the interaction of the larger societal and environmental frameworks that prefigure and shape the events and processes of disaster, aftermath and recovery represent the central issue of my focus at present. In a sense, disasters may represent a set of effects that "arise from the 'instantiation' in nature of certain kinds of social relations" (Harvey 1996: 200). In that sense, disasters can provide insights into the nature of specific sociocultural forms of adaptation, including productive forms, social relations, and ideological constructions. Writ large, if sufficiently attended to, disasters contain important messages regarding the appropriateness and adequacy of the organization of both social and human–environment relations.

EPILOGUE

WHERE WE HAVE BEEN AND WHERE WE MIGHT GO

Putting the elephant together, blowing soap bubbles, and having singular insights

E. L. Quarantelli

This last chapter presents my overall reaction to what was said in this volume as well as elsewhere about the question of conceptualizing what is a disaster. The subtitle given to this chapter is my effort to dramatize that I envision the conceptual differences which exist as mostly reflecting basic but diverse assumptions about the nature of reality and knowledge (on philosophical issues in general, see Bunnin and Tsui-James 1996; Bynagle 1997). As I see it, there are three fundamentally different images of the phenomena. Some researchers operate with a model analogous to the parable of the blind men touching an elephant. This approach is very widespread in social science theorizing. It has been explicitly advanced for the disaster area (e.g., Short 1989). The basic assumption in this form of philosophical realism is that there exists an elephant (i.e., a disaster) out there whose parts and whole can be eventually pieced together. Then there is the dissimilar image of other researchers with a more social constructionist view of phenomena. They see researchers separately blowing soap bubbles that vary in hue, size and duration. They can agree that the bubbles (i.e., concepts of disasters) are socially created, but otherwise disagree because everything about the phenomena is constructed. Then there are still other students of disasters who assume in a solipsistic way that the actor's perspective is paramount. As such, everyone can provide a singular subjective insight into the phenomena that others might not see, so victim views of disasters are as equally valid if not more so than a researcher's view. As readers will note, this mixed metaphoric imagery is the implicit background for what I will say, although I will only explicitly return to these ideas at the very end of the chapter.

I now turn to my specific remarks that are organized around four general observations. The first two have to do with what was set forth by the other writers and also in the theoretical disaster literature generally. That is now all in the past. The second two have to do with what might be done in the future in trying to answer the question: For research purposes, what is a disaster? In short, as a field where might we go from here? To some, my set of remarks might seem to be a brief written for a devil's advocate. That is a partly correct perception. My provocation is deliberate. It stems from my belief that future scholars must address key issues embedded in my general observations about past efforts to deal with the conceptual question of what is a disaster.

As I see it, (1) the various authors in this volume, reflecting much of the existing theoretical literature, while similar in their views along certain lines, nevertheless show significant differences in their answers to the basic question. (2) However, this overall exercise of asking what is a disaster, was quite useful and worthwhile for all concerned including myself and an activity to be further encouraged. (3) Furthermore, this past work does suggest in my mind a major path future students of the problem might follow, namely, developing a *typology of crisis situations* (within which disasters are only one type). (4) Additionally, it seems to me that we should at least consider whether current efforts to develop an answer to the conceptual question might be too reformistic, and that what is needed is more revolutionary, the creation of a *new paradigm* for disaster research.

This is all based on the assumption that different answers to the basic question makes a difference. If it does not, then this exercise should never have been launched in the first place. However, I more than agree with Turner who notes that: "of course there can be no theory and no scientific investigation without classification, as Herbert Blumer (1931) reminded us long ago" (1989: 265). Or, as even more precisely stated by Robert Merton a long time ago:

> concepts constitute the definitions (or prescriptions) of what is to be observed; they are the variables between which empirical relationships are to be sought . . . it is . . . one function of conceptual clarification to make explicit the character of the data subsumed under a given concept . . . our conceptual language tends to fix our perceptions and derivatively, our thought and behavior. The concept defines the situations, and the research worker responds accordingly . . . conceptual clarification . . . makes clear just what the research worker is doing when he deals with conceptualized data. He draws different consequences for empirical research as his conceptual apparatus changes.
>
> (Merton 1945: 465, 466, 467)

A recent article by Dove and Kahn (1995) well illustrates this. Without showing any awareness of the existing conceptual dispute, they discuss two "competing constructions of calamity" that they empirically apply to a 1991 cyclone that hit Bangladesh. In my view, they clearly show that the label which is applied (a naturalistic versus a political one) makes a difference in what is observed and identified as important with respect to both the characteristics of and conditions for that disaster.

HOW AND WHY DIFFERENCES ARE FOUND

A US Supreme Court justice supposedly said that: "I cannot define pornography, but I know it when I see it" (paralleling St Augustine's alleged statement that "If no one asks me I know what time is, but if someone asks I do not know what it is"). My general impression is that most disaster researchers have the same attitude about the defining concept of their field. Now some authors in this volume do explicitly venture definitions of disasters, but, as I see it, do not set forth completely compatible formulations. Still others either dismiss the basic question or suggest the issue ought to be differently addressed (outside this volume see, for example, Bates and Peacock 1989; Drabek 1989b). I have no problem with these last points of view as such. They can be legitimate positions. However, the very actions involved imply at least an intuitive identification of a phenomenon called "disaster." Even those who dismiss the question must have some image of disaster in mind, for otherwise how could they say that asking about X (disaster) is not the best way to proceed? What is the nature of X that is being dismissed as less important than some other perspective? Likewise, if the question ought to be asked in a different way, that way has to have some point to which it is a reaction from. Different from or away from what? In my view, the assumption in both the argument for dismissal or going elsewhere, whether recognized or not, is "I cannot define disaster, but I know it when I see it."

I am not certain that the American justice would be willing to accept the on-the-face-of-it argument from a lawyer pleading a case before the court. However, I am willing to do so for purposes of discussion here, primarily because I think it is the way most students of disasters actually operate. Whatever formal definitions they might have, *in practice* they use the label at most as a gross sensitizing concept (see Blumer 1954, 1969, who argues that such conceptions call attention to "something" by giving it a label, and that the labels act not as prescriptive guides but as suggestions of the direction in which to look). An inevitable outcome of so proceeding is that different conceptions are implicitly, if not explicitly, advanced. This should not be surprising. When explicit formal definitions are advanced they mostly differ, so implicit views are likely to be even more different. I use this as a starting point to examine what seems to be done when scholars are thinking about the question, and why at one level (basic assumptions aside) they produce conceptions that are both structurally and substantively different.

As indicated in my earlier remarks, when in 1992 I launched the enterprise that eventually turned into the preceding essays and semi-dialogues, it was not clear to me how much agreement would emerge. However, I clearly expected some degree of disagreement otherwise it would not have made much sense for me to initiate the exercise in the first place. Yet, even at the time of the meeting at the Sorbonne, it was never my thinking that there would emerge overwhelming convergence on answering the basic question posed: For research purposes, what is a disaster?

What have others concluded when they have looked at the answers of the different authors in this volume? Both Hewitt and Perry note some similarities in points discussed in the papers to which each reacts, although this conclusion is seemingly based more on the structure of the arguments advanced rather than substantive content. For Hewitt, there are commonalities in what is left out as well as in what is addressed. Thus, he observes that the authors to which he is reacting are less concerned with defining disaster than they are with the preoccupations and paradigms that shape how the phenomena are approached. In particular, he argues that they are relatively similar in that they all come out of the relationship between Western governmental and professional views of the crises that have to be dealt with. Hewitt then goes on in a mostly postmodernistic vein to elaborate important dimensions that he thought had been ignored by all the authors in the first round. Making as good a case as I think is possible for his position, in so doing, he is indicating to me that he is conceptualizing disaster in a different way than many others.

Perry, in a very sophisticated paper, sees even more similarities among the authors to which he reacts. He notes that there is agreement on seeing disasters as socially defined occasions and as involving radically changed behavior to meet a crisis. In his view too, all the authors in the second round see disasters within a social change perspective and view the concept as a multidimensional one. He also notes most of the essayists "tell us more *about* disasters, rather than answering the query: What is one?" Nevertheless, Perry in his reaction chapter does also grant that there are differences which arise "from inductively stimulated reconceptualization over time, while other variations reflect different disciplinary or theoretical frameworks, or units of analysis orientations of the social scientists." I agree.

On the whole, I would not dispute the specific assessments, as far as they go, of either Hewitt or Perry, including that there are some commonalities. However, I do not see them as going far enough. Thus, I see far more differences than they do, which may partly reflect the probability that they are far more willing to be polite or less confrontational than I am. At any rate, from my perspective, there are significant differences among the authors (and not incidentally, including and notably between Hewitt and Perry) in several ways, both in the structure of their arguments as well as substantively. For one, there are explicit/implicit disagreements on certain issues, although maybe some can be finessed as Perry ably does with the social change/social problem controversy. It is notable too that most of the authors vary considerably in what they suggest should be done in the future. At the extremes, they range from those who think knowledge can be obtained incrementally through continuing traditional *objective* empirical studies, to those who advocate obtaining understanding through adopting a qualitatively different *subjective* approach that would see disasters mostly from the perspective of the victims. Thus, to me, and simply as an example, Kroll-Smith/Gunter and Kreps are operating in different intellectual worlds on their views about future work in the area.

In addition, as Hewitt in particular notes, and which I think is also true of the authors in the second round reacted to by Perry, there are major differences in the sense of the ignoring or not discussing issues which others consider important. For example, few of the essayists, except Perry and Rosenthal, directly and explicitly address the question of whether future disasters will differ in consequential ways from past ones and how this could affect the conceptualization process. As I will note later, slighting or ignoring issues others consider important or crucial is also a way of indicating that one's own conception is a better way of defining a disaster.

The classification of conflict situations

This said, let me turn now to discussing part of what is involved behind the differences that both implicitly and explicitly emerge. With a few notable exceptions, most authors in this volume and reflecting the theoretical literature generally, do not in any explicit fashion advance or apply specific operational criteria that might be used in any classification attempt. In what follows, I will discuss *conflict* situations, crises that allow me to illustrate at least one major way typically used to approach the classificatory problem. (The chapter by Gilbert well links military war needs with pioneer studies of civilian disasters, but his goal is a historical reconstruction of the development of the field and not the classificatory problem per se). My intent here is not to definitively settle the conceptual issue of the placement of conflict phenomena. It is mostly to exemplify a way we all use in addressing whether certain occurrences should be included or excluded under the conceptual rubric of "disaster." As such, equally as important as the distinctive characteristics of the behaviors I discuss is that I will be illustrating what all of us actually take into account in our categorizing into a classificatory scheme.

What might be used to include or exclude any happenings under what else we put under the label of disaster? Indisputable candidates are empirical data, theoretical notions, and logic analyses, or mixtures of them. The last might be particularly useful given the "chicken or the egg first" dilemma in trying to identify disasters in the first place and, at the same time, establishing the characteristics such phenomena exhibit. (In earlier writings, I have noted that definitions and concepts while related do not have the same referent, but this distinction is not herein anywhere further discussed.)

The question of whether conflict occurrences ought to be treated as "disasters" has plagued the field of disaster studies from its beginnings. I remember staff discussions at the National Opinion Research Center in the early 1950s on whether riots, etc. should conceptually be treated in the same category as the behaviors seen after a tornado or mine explosion. The question was never formally resolved, but the leaning was to separate the two phenomena. In part, this was because on the basis of the earliest field studies of behavior in sudden natural/technological crises, it was concluded there was neither the personal nor social disorganization that common sense and stereotypic notions implied should have been present. In contrast, a revolution or mob behavior seemed to exhibit more disorganization. Even I might now question this line of thinking, but it prevailed and is a reason many pioneers in the area initially separated out conflict situations from disaster occasions.

Whether for that or other better reasons, it is a fact that in the last four decades, extremely few self-styled disaster researchers have personally studied *both* conflicts and disasters or attempted any systematic comparison of the behavior in the two settings. Certainly this has not been from a lack of conflicts for examination; there clearly are far more of them than disasters for study purposes. Also, there is a considerable body of empirical and theoretical literature that deals with conflict situations (over decades, see bibliographies such as by Morrison and Hornbeck (1976) and Zimmermann (1983); and recent collective behavior works such as by Oliver (1989), Marx and McAdam (1994), McAdam, Tarrow, and Tilly (1996), Melucci (1996) or the journal *Mobilization*). So anyone wanting to make a systematic comparison cannot argue that relevant data and ideas for such a purpose are absent.

Among the few disaster researchers who at least looked at conflict situations were some at the Disaster Research Center (DRC). In the early 1960s, DRC did field studies of personal and organizational behavior in student and ghetto riots. It was semi-explicit in the thinking of those involved that a comparison was being made between behaviors found in riot and disaster situations. Empirically, differences were found, especially in the ghetto riots. Fire and police departments, for example, planned for and responded to the two situations in distinctively different ways (e.g., Kreps 1973; Wenger 1973; Weller 1974; Warheit 1996). These ranged from the different training undertaken to what factors entered into organizational decision-making (e.g., explicit political considerations were very important in riot settings). To this day, many departments in the United States continue to have different planning and standard operating procedures for the two kinds of crises.

Hospital operations were markedly different in the two settings (Quarantelli 1970a, 1983; Tierney and Taylor 1977). Thus, in riots, because of curfews and the violence in the streets, hospitals were often caught with using only the shift on duty instead of the three frequently available in disaster occasions. The flow of casualties into hospitals peaked relatively quickly and then sharply fell off in disaster occasions, whereas a far more erratic pattern of inflows and multiple peaks occurred in riots. Medical personnel were sometimes the direct objects of attack in a riot. This never happened in a disaster. Also, unlike in disasters direct confrontations occurred between responding personnel with prime responsibility for attribution of criminal conduct and for those with only health concerns; this can affect the nature and duration of the search and rescue efforts (this also occurred in both the recent World Trade Center and Oklahoma City bombings). Unheard of in disasters, victims in ghetto riots fought with law personnel in emergency operating rooms or simply refused medical treatment. A physician interviewed by DRC who had spent years in Lebanon during the civil strife, said that emergency medical activities in Detroit during the 1960s rioting were far more similar to what he had encountered overseas, than what he had experienced in tornadoes, floods, and non-terroristic explosions in this country.

DRC also discovered that even behavioral phenomena that at one level seemed the same, manifested different characteristics when looked at closely. For instance, what about the looting in the two crisis situations? In disaster occasions, looting is very rare if nonexistent, covertly undertaken, done by isolated individuals or pairs, socially defined as reprehensible, and mostly takes place at targets of opportunity. In contrast, at least in the ghetto riots, looting was both common and widespread, overtly undertaken, often collectively by many primary groups including families, was socially supported, and mostly was selectively done at targeted locations or places (Dynes and Quarantelli 1968; Quarantelli and Dynes 1970). More recent observations of looting in the Los Angeles riot of 1992 (Tierney 1994) are consistent with earlier observed patterns of massiveness, overtness, and selectivity.

The point of the above discussion is that the issue of classificatory inclusion or exclusion can partly be addressed at the empirical level. Are the phenomena that appear similar or are they dissimilar? My reading of the data is that there are significant differences at the individual and the organizational behavioral levels. However, even as I myself have written elsewhere, we can use empirical data only so far in categorization efforts (see Quarantelli 1993a) because there is the "chicken and the egg" problem here in the use of any empirical information. Nevertheless, since such empirical findings as exist support making a distinction, they back those who distinguish riots from disasters.

At a more abstract or theoretical level are a very few attempts at direct systematic comparison (for example, see Warheit 1968; Weller 1972; Rosenthal, 't Hart, and Charles 1989). In 1993, I did review research findings that allowed a general comparison between behavior in natural/technological disasters and in riots and civil disturbances in the United States. Behaviors at the individual, organizational, and community levels in the preimpact, impact, and post-impact stages of both kinds of situations were comparatively examined. Overall, while there were some behavioral similarities, especially at the organizational level, I found there were far more differences, some of a rather marked nature. For example, when disasters occur, individuals actively react and with a prosocial mode; there is far more variability in riots with antisocial behavior frequently surfacing and a markedly different pattern of looting in the two settings as previously noted. Also, while a disaster experience is a memorable one, and there are differential short-run mental health effects, there does not appear to be too many lasting behavioral consequences; riots seem to leave more residues. Similarly, there is somewhat more likelihood for organizational changes after riots than after disasters. At the community level, disasters generate massive convergence behavior; this is far less true of riots. While there are selective longer-run outcomes and changes in impacted communities, the impact is less in typical disasters than riots. Thus, our comparative summary of a range of empirical data supports conceptualizing at least some major conflict situations in a different way than disasters (Quarantelli 1993a).

Finally, at a primarily logical level, it seems to make a difference if some phenomena are of a *consensual* nature, and some are of a *conflict* nature (e.g., the social movement literature has also found it useful to distinguish between consensus and conflict type movements, see Lofland 1989; McCarthy and Wolfson 1992; Schwartz and Paul 1992). Disasters are consensus occasions while riots are conflict situations. By consensus, is meant that those participating in the situation are generally in agreement that the crisis should be brought to a halt. This does not mean that there is no conflict in disaster occasions (just the opposite has long ago been documented, see Dynes and Quarantelli 1976). However, what is absent is a division into competing groups or factions, at least one of whom is interested in making the situation worst for other parties. The latter can be seen in wars, revolutions, ethnic/religious clashes (like in contemporary Bosnia, Sri Lanka, Northern Ireland), riots, terrorist attacks including the gas poisoning in the Tokyo subway, and sabotage efforts.

In citing these findings and arguments, I am making two points. First, there are empirical research results, comparative findings, and logical analyses that indicate behavior differs in riots and natural/technological disasters. Consequently, anyone who does not want to draw such a conceptual distinction, will have to explain away the differences found. I do not think that will be easy without comparative data. (Kreps does have a current study systematically comparing historical data on initial organizational responses in riot and disaster settings that should surface whatever similarities and differences there are.) Second, and more important, is that it is probably the use of such mixtures of research observations, implicit comparisons, and logical arguments that actually lead almost all researchers to exclude and include whatever they do under the "disaster" label. It is very rarely the formal definitions they use. As I have just illustrated, it is my mixing of a variety of factors that have led me to exclude conflict situations from being usefully classified under the label of "disasters."

Finally, if my analysis is correct, it is clear, at least to me, why it will be a long time before we get much agreement about what is a disaster. We all use somewhat different mixtures of empirical observations, theoretical leanings, and logic to do our classifications. So, whatever our formal definitions, the "I know it when I see it" view necessarily leads to considerable variation in what is and is not classified as a "disaster." The obvious is only obvious to those who already "know" what they will see. Or as Dombrowsky (1995b: 242) put it: "we see what we want to see."

Now some might see my observation that there is and will continue to be different conceptions of disasters as depressing, and perhaps indicating that there should be an abandonment of the exercise of trying to develop more agreement. However, I would think that for social scientists who think their work valuable, the existing state of affairs would be a challenge, something worthwhile examining even more. If answers in research were easy, it would not be very interesting or much fun.

VALUE OF THE EXERCISE

From the start of this exercise, while I did not expect and did not see a basic convergence emerging, it was my expectation that asking leading disaster researchers to make their views explicit would somehow advance thinking on the topic. Along this line, I do believe the disaster research community now has a much better understanding of what is involved in trying to answer the question as to what is a disaster, and why advancing answers is important. In my view, even the differences that have surfaced and the issues that have been ignored or slighted have moved the field considerably forward in thinking about the phenomena.

My strong feeling is that the exercise in this volume has accomplished much, particularly along three lines. (1) From a historical viewpoint this is the first time that there has been as much time and systematic attention paid to the basic question, involving a dozen researchers from six disciplines and six societies. (2) The proponents of different views have had to be far more explicit about their ideas than they ever previously had to be. And (3), new ideas, at least to me, were expressed.

The systematic and large-scale approach

For the first time in disaster studies, a major effort was made to pull together in one place as many as possible of the various conceptual views that existed. Researchers from around the world participated. Also, a semi-dialogue format was used. To the extent there were previous efforts, they were far more limited in scope, range, and intent, usually had no participants other than Americans, and did not allow any printed intellectual exchange. Whatever one may think about the different essays or remarks of any of the authors in this volume, what has been produced will be a benchmark in later social science disaster research. Future theoretical efforts will be evaluated to the extent to which they improve or extend what has been done in this volume.

The explicitness of views

Undoubtedly, new thoughts came to some of the authors in writing their essays. However, I think that most really made explicit what they had already previously implicitly believed. Also, to some extent, because of the semi-dialogue nature of the volume, proponents of different conceptions had to clarify and elaborate their particular views far beyond what they had previously done. While the dialogue was not as intensive as I would have liked, nevertheless, many had to discuss somewhat the implicit/explicit criticisms or questioning of their particular perspective. Making explicit what had been implicit can only be good in the development of a field of study.

The expression of new ideas

To me at least, some new ideas about the problem of defining and conceptualizing disasters have come to the fore. Since different readers will undoubtedly find as "old hat" what I thought was "new," and vice versa, I will forego identifying what was not part of my thinking before. However, since I think that any reader will learn something new from the previous chapters, I do consider that as partly making this exercise worthwhile.

For these three reasons I see this volume as showing intellectual progress. Only out of explicit and clashing writings on the problem and trying to meet criticisms will the field move toward a necessary rough agreement on its basic concept. Some authors were unnecessarily polite, leaving their criticisms unstated, which was not a contribution to an exchange of views. Fortunately, others dealt with what their colleagues said. In several cases the exchange was spirited, but since such clashes are indicative of different basic perspectives, fundamental disagreement on how to view disasters was surfaced.

To some readers of this collection of essays, the volume might seem to have only further complicated the picture of where the field stands. Perhaps that is so. But as Aristotle wrote a long time ago, the emergence of wisdom requires an initial period of confusion. Undergoing that last state myself, this overall exercise forced me to think more intensively about three notions

I had toyed with in my earlier work. Because of what others have written in this volume, I now better understand what I previously thought. This includes: (1) a reinforcement of my belief of a need to continually refine the key constructs of a field, such as the concept of disaster; (2) a strengthening of my thinking that there is a need to develop a typology of disasters; and (3) a conviction that there is an even greater necessity now for a typology of crisis situations (of which disasters would be only one type). Since I will look at the first two primarily in terms of the past, I will discuss them in this section of the chapter, whereas the last point, because it deals mostly with the future, will be examined in the following section.

A refinement of the concept of disaster

There are many ways in which the concept of disaster could be refined. One that I have suggested for some time is that disaster researchers ought to disentangle disasters from being solely linked to physical agents or the physical environment. I see reinforcement of this view implicitly if not explicitly in the comments of some authors in this volume, although they cannot be held responsible for my interpretations of them. Let me give three examples.

Stallings in his reaction note proposes that taxonomies be developed on the basis of exceptions to routines in the major institutional sectors of society. In my view that approach has merit because it allows a conceptualization that goes far beyond rooting disasters in just a natural/technological agent risk base. His example is that the Swiss banking community might be faced with an economic "disaster" because of its activities in handling German assets during World War II. Whatever the merits of the example used, his point about focusing on institutional sectors to develop types seems worthwhile pursuing by anyone who wants to divorce disaster occasions as occurring only from technological risks and physical hazards.

Rosenthal, too, suggests that there are limits in traditional conceptions of disaster. In particular, he argues they use a very narrow time–space setting in looking at the conditions, characteristics, or consequences of such happenings. His position is that disasters before, during, and after their occurrences, happen in a very complex world of linkages, chains, and processes. As such, that must be taken into account in conceptualizing disasters. I mention his ideas because generally they seem to fit with our notion, discussed later, that conceptually we should go from an *agent* focus to a *response* focus orientation. Again, whatever the substantive merits of the view that disasters need to be looked at in terms of their conditions, characteristics, or consequences, his view is a good starting point for anyone who wants to conceptually go beyond sudden natural/technological agents.

Perry takes the general idea of possible reconceptualization even further. He writes that it might be possible through the use of taxonomies to include a variety of phenomena under one general or generic definition. In his discussion, he notes that it might be possible to develop typologies along both phenotypical and genotypical lines. If this were done, I would think that physical agents might become of secondary importance for conceptual purposes, something with which I agree (and which also seems consistent with what Oliver-Smith has written not only in this volume but also elsewhere, as in his review article of 1996).

The views expressed above, at least as I have interpreted them, are consistent with my greatly ignored Presidential address to the ISA Research Committee on Disasters more than a decade ago when I said we should follow the typical course in the development of scientific concepts. I wrote that:

> Pioneers start out with common sense or everyday ideas. Those that follow reformulate the basic ideas so that in time key concepts have more meaning within the scientific discipline than they have in popular discourse. In fact, the scientific concepts and ideas eventually get to be rather distant from common-sense notions (e.g., that color is an integral characteristic of physical objects, or that heavier objects fall faster than lighter ones, or conversely, that bats, whales, and human beings share many characteristics in common as mammals).
>
> (1987b: 27)

I then said that I thought we needed to follow the lead provided by biology that draws a distinction between phenotypes and genotypes. Since my days of taking courses in college biology, I have always been impressed by the notion of *phenotypical* (surface or manifest characteristics) and *genotypical* (common nonvisible factors) concepts. As such, I would not be disturbed if our concept of disaster is eventually genotypical. Furthermore, continuing to cite this source with which I usually agree:

> such a concept may violate common sense. Assuming . . . the consensus-conflict distinction we drew earlier, there might be a conceptual distinction between, e.g., a plane crash generated by a terrorist bomb and one by an engine malfunction. So what? The issue is whether we can learn and study more that way, rather than whether distinctions violate common sense . . . It is on the research payoff that the judgment is to be made. We will have made tremendous progress when as researchers we will be able to talk meaningfully of type X or type Y disasters rather than of hurricanes or chemical poisonings; in fact, we would not use such terms for research purposes as tornadoes or explosions, because different ones of such social occasions would be classified as Types A, B, C, etc.

> disasters—the A, B, Cs, etc. eventually being given labels for which no common sense or everyday terms presently exist.
>
> (Quarantelli 1987b: 28)

Or, as a participant noted in a recent electronic discussion group (Griffin 1995), should we distinguish between wildfires produced by lightning and those by tourists who fail to extinguish cigarettes? My answer is, Yes, if making such a distinction will lead to better understanding and knowledge of the social behavior involved.

Many with postmodernistic orientations will of course find this approach heading in exactly the wrong direction. Similarly there will be those who will raise questions about the development of jargon. In my view, they both miss the point. We need to move away from just the views of victims or the visual implications of everyday words. Any jargon, whether in baseball, chess, popular music, or science, makes for precision rather than the reverse, as common sense might imply. As such, in the good sense of the term, we need *more* specialized jargon in disaster research. Our continuing dependence on the jargon inherent in everyday or popular speech continues to blind us to other more useful ways of looking at "disasters".

In that respect, having studied too long ago as a graduate student with O. D. Duncan, I have been long aware of the POET framework that involves four basic concepts—population, organization, environment, and technology (1964)—which in some formulations attribute the greatest importance to the environment (Schwab 1993: 38). I have been surprised that disaster researchers who want to keep the physical environment central to their conception of disasters have not used the POET paradigm. There is much in that formulation that could be used as a larger framework to bolster their point of view. From my perspective there is far too much reification of the key concepts. I also am not imaginative enough to see how abstract intellectual tools, i.e., concepts, can interact (e.g., how does environment interact with technology; but see counter arguments in Catton and Dunlap (1978), Buttel (1987), and Freudenburg and Gramling (1989)). Nevertheless, I can see where others might feast on the POET paradigm and I assume that is what is involved in Short's statement that:

> isolation of the social from the physical environment has proved to be a hazardous undertaking . . . the ability even to conceptualize the potential of . . . disasters requires recognition of the human species' dependence on the biosphere.
>
> (1989: 402)

Or as Dunlap and Catton more bluntly state it for sociologists:

> The Durkheimian legacy suggested that the physical environment should be ignored, while the Weberian legacy suggested that it would be ignored, for it was deemed unimportant in social life.
> (1983: 118)

To some this explains why as a sociologist I am on the wrong track! The "obvious" is "obvious."

However, I see our responsibility as researchers is in the longer run payoff. It is said that Benjamin Franklin was once criticized for flying kites in thunderstorms. Supposedly he was asked what could possibly be the practical outcome of what he was doing, which of course missed his objective of studying the general nature of electricity. Franklin answered the question of the value of his work with a question of his own. He asked: Who have saved more lives—the carpenters who build better lifeboats or the astronomers who first studied the distant stars that eventually contributed to better ship navigation? We ought to be astronomers. Carpenters are needed, but as researchers we have a different responsibility, the same as astronomers. Such a path will take us away from the everyday familiar, but it is the one that will have the greatest payoff in the long run, even in practical terms. In some ways I am arguing that we should root our research as researchers in the theoretical/conceptual area rather than building on applied or practical questions. Although I do not know if he would agree with me on the position just expressed, I should note that the article by Porfiriev has an excellent discussion of how research in the area can be rooted in two different approaches, what he calls the applied/pragmatic one and a theoretical/conceptual one, and the difference that makes in any research undertaken, whether in the former Soviet Union, Russia, or any place else.

I do recognize that valid research findings are not necessarily, and in fact seldom immediately, useful. However, we should remember that another major finding of disaster studies is that organizations which delay responses and assess the overall situation usually are far more efficient and effective in the crisis than those who plunge ahead in terms of what may seem a crucial problem that demands immediate attention. To me, there is a lesson in that about what disaster research social scientists should be doing; instead of simply reacting to the "obvious" and plunging ahead, they should step back and take their basic research ideas from the abstract theories and models in the various disciplines of which they are an intellectual member.

The need for a typology of disasters

Apart from clarifying the concept itself, another past position of mine has been that the study of disasters also needs a typology of some kind. What some authors in this volume have written has strengthened my belief that this is the right tack to take. The need for such a typology is well argued by

Perry in his reaction paper. It is even better illustrated in detail by Dynes in his primarily typological article, although he pulls back a little from the idea in his reaction note.

Of course, in my view, the field has long been misled by a phenotypical distinction drawn from common sense. Thus, consistent with a move in a genotypical direction, I would restate that, as far as I am concerned, it is not worthwhile for research purposes to further pursue a typological distinction between natural and technological disasters. As indicated in our later citations of Erikson and Picou, not everyone agrees with this. However, as can be seen in many publications by myself (e.g., Quarantelli 1982, 1987b, 1991, 1992a, 1993b, 1995a) and others (e.g., Tierney 1980: Wijkman and Timberlake 1984; Bolton 1986; Bolin 1988; Mitchell 1990; Rochford and Blocker 1991; Towfighi 1991), very extensive empirical studies, theoretical ideas, and logical analyses have been used to challenge the supposed difference. Given that huge literature, there is little purpose in restating its criticisms here. That most scholars in the area have taken them into account seems to be illustrated by the fact that the great majority of authors in this volume do not allude to or attempt to make such a differentiation. It is also not insignificant that the distinction has been increasingly abandoned in much emergency management operations around the world (although, as Waugh 1997 correctly points out, this may not be an unmixed political blessing along with the professionalization of the emergency management field).

Although I stopped using the natural/technological disaster distinction long ago, I have always felt that there are other features that might be used to start to distinguish certain categories of disasters. For example, some threats such as hurricanes and most nuclear plant radiation accidents allow warnings, while others such as earthquakes and most chemical explosions do not provide forewarning. Some impacts are very localized (many tornadoes, explosions), while others are very diffuse (most river floods, hurricanes, and many hazardous spills). However, no single dimension is enough on which to base a typology. As Perry correctly notes, most disaster researchers think of disasters in multidimensional terms. I do not see reasons to challenge that view. So, to state it very specifically: What is needed for disaster research is a typology based on general dimensions that not only cuts across different disaster agents, but also the same disaster agent. As far as I know no such typology has ever been advanced by anyone. In this volume the article by Dynes starts in what I consider the right direction, and it is possible Perry, Rosenthal and Stallings might also accept the direction just suggested. I will leave it with saying we have a long way to go yet.

To me, this assessment is reinforced by the fact that no matter how it might be ignored by some, the basic question is not going to go away. As in my view, Dombrowsky correctly notes the question is asked over and over again, although not everyone recognizes that they are so doing. However, as a perusal of the literature will show, even just since the writing for this volume started there continues to be an asking of the conceptual question. Of course, most of those who ask are concerned with obtaining an answer for administrative or operational purposes rather than for research use (but see Porfiriev in this volume or Britton 1987, who earlier look at possible links between the two areas).

However, Richardson (1994) did ask the question for research purposes. Operating within an industrial crisis management framework, he sets forth four types of crises: ecosystemic disasters, socio-technical disasters, business-economic failures, and sociopathic attacks (in his one passing remark alluding to natural disasters, he says his interest is only in organizationally induced disasters). To an extent, he builds on earlier formulations about industrial crises set forth by Shrivastava, Mitroff, Miller, and Miglani (1988), and indirectly by others (e.g., Perrow 1984). In the same tradition, is Kovoor-Misra (1995) who advances a sevenfold typology of crises, namely technical ones (e.g., leaks), political (e.g., negative publicity), legal (e.g., liability), economic (e.g., bankruptcies), ethical (e.g., corruption), human and social ones (e.g., terrorism) as well as natural disasters (e.g., floods). However, some in this volume, such as Hewitt and even myself, might see a problem in developing ideas about disasters that are drawn mostly from crisis responses of governmental and business bureaucracies.

Also, as I noted in my original epilogue (Quarantelli 1995b) to the journal issue, Erikson with seemingly more of an understanding rather than research knowledge goal published *A New Species of Trouble, Explorations in Disaster, Trauma and Community*. In this book he states that the world is moving into a dystopic era where new problems that traumatize communities:

> are seen as having [*sic*] produced by human hands, they involve some form of toxic contaminant, and they blur the line we have been in the habit of drawing between the acute and the chronic.
> (1994: 22)

In a footnote he presents a fourfold typology of this and related phenomena with technological and natural along one dimension and toxic and non-toxic along the other (ibid.: 246), but no generic definition of disasters is explicitly presented. However, it remains to be seen how disaster researchers will respond to a 1994 publication which cites no empirical studies of natural disasters after 1957 and no general theoretical work after Barton's book of 1969. Even on technological disasters, it cites mostly earlier work on Three Mile Island, almost none of the later more sophisticated studies, and only alludes to Chernobyl in passing (see an annotated bibliography on the massive literature in both English and Russian sources in Quarantelli and Mozgovaya (1994)). Perhaps this is an illustration of what Kroll-Smith/Gunter mention in their earlier reaction note, that when scholars with a particular approach limit their literature citations to one another, an orthodoxy is likely to emerge.

That theoretical issues seemingly resolved tend to come back is illustrated by an even more recent publication by Picou and his colleagues (1997). It uses the *Exxon Valdez* oil spill to also argue that technological disasters are fundamentally different from natural disasters, and that such happenings should be viewed within a social problem framework. In a slightly earlier publication, it is said:

> Over the last 25 years, the concept of disaster has been discussed, debated, and empirically studied in relation to an increasing variety of catastrophic events. As a result, the concept of "technological disaster" has emerged, bringing attention to the unique qualities and impacts of events that arise from failures of technology.
>
> (Picou and Gill 1996: 879)

Many other scholars beyond myself would undoubtedly challenge the accuracy of the supposed historical trend and conclusion, as well as the attribution of uniqueness to technological disasters, but the stated position is unambiguous.

Finally, as a last example, is Alexander, who while paying almost exclusive attention to natural disasters (just the converse of Richardson) nevertheless writes:

> The distinction between natural hazards or disasters and their manmade (or technological) counterparts is often difficult to sustain (1993: xv) . . . we are dealing with a physical event which makes an impact on human beings and their environment . . . a **natural disaster** can be defined as some rapid, instantaneous or profound impact of the natural environment upon the socio-economic system.
>
> (1993: 4)

In a later publication Alexander (1995) also states that the fundamental dimensions of extreme geophysical events are time, space, magnitude, and intensity, which along with his other statements some might see as a form of geographic determinism, even if that is almost certainly not intended by the author.

The above, atypical only in that the views expressed are very explicit, are but examples of what is constantly appearing. As such, the works by Richardson, Kovoor-Misra, Erikson, Picou, and Alexander (along with others previously mentioned, for example, Drabek (1989b)) show that there continues to be far from agreement on what is a disaster, which is also my reading of the authors in this volume.

We can now all treat what has been done in this volume, and in the prior literature, as the past. However, we can do little about what has already occurred. Therefore, my preference at this point is to start suggesting what might be dealt with by future researchers interested in addressing theoretical aspects of disasters. The various authors in this volume have had their say and, almost all, also a rebuttal opportunity. Readers can make their own judgments on how well the essayists met the initial charge that I provided, the value of whatever they said about the question generally and one another, and the validity and relevance of my previous comments about them.

TOWARD A TYPOLOGY OF CRISIS SITUATIONS

As such, let me turn away from primarily a focus on the authors in this volume and start indicating what all of us might do in the future. With respect to that I want to suggest a possible fruitful way of getting more movement on the basic question, that is, through the development of a typology of *crisis situations* (of which disasters are only one type). The idea is not a new one. Barton in his pioneering work decades ago presented a formulation that used four dimensions to build a typology (scope of the impact, speed of the onset, duration of the impact itself, and social preparedness (1969: 41)). This led him to set forth a typology of what he called collective stress situations that included as examples the sudden death of a head of state, air bombing campaigns, plant explosions, community tornadoes and floods, ghetto riots, national depressions, famines, genocides, pogroms, mining "ghost towns," areas subject to malaria, the status deprivation of untouchables and the poverty of the lowest income groups, long-standing slums, loss of support by a religious sect, economically backward areas, and being a chronic minority party. Unfortunately this typology, which makes far more sense when seen in the larger context of his full remarks, was not built upon by anyone.

However, the use of typologies in the social sciences is an old activity and continues (McKinney 1966; Bailey 1992). Even the general value of their use in connection with disasters has been previously discussed. For example, a methodologist (and not a disaster researcher) has written:

> Despite some problems, which generally plague all of social research, taxonomy promises large benefits for disaster research. It not only aids in cataloguing, comparison, and research genesis (in its theoretical mode), but also similarity, thus, facilitating explanation and prediction (in its conjoint mode). Its empirical mode is conducive to computer-aided generation of taxonomies, what might be termed **grounded taxonomy**.

After suggesting three strategies for typological formulations, he concludes:

> A final point to make is that the suggested exercises in typological analysis promise reasonable yields while costing relatively little. Typological analysis is amazingly complementary to other forms of analysis. Constructing typologies generally does not preclude other analyses, and is generally not particularly expensive nor time-consuming relative to other methods. Rather than being an expensive luxury, typological analysis of disasters is instead a valuable foundation and complement for other forms of analysis, and this valuable tool should not be neglected.
>
> (Bailey 1989: 429–430)

Now, while Bailey is primarily discussing developing typologies of disaster phenomena, I think his ideas are equally applicable to a typology of crises.

The related question of creating taxonomies was partly addressed nearly a decade ago in an issue of the *International Journal of Mass Emergencies and Disasters*, edited by Kreps (1989c) who also contributed substantially to the discussion. For my purposes here, I will ignore that the terms, taxonomies and typologies, have different referents. The first has to do with principles of classification, the latter with the end product of a classification (Bailey 1992). Building either one, while a related activity, is not the same process. Some might argue that creating typologies cannot be done without developing a prior taxonomy, a fatal problem for my argument if true, but I do not see it as a valid point for reasons suggested later.

However, I more than grant that there is a great danger in creating typologies of any kind. This is that they inherently give the impression that because the scholar is producing "types," something is truly being accomplished. Often that is nothing more than simply attaching labels to projected cells, an endeavor that is easy and can be done almost endlessly. This can become very seductive. However, the endeavor will not be fruitful unless the types can be used to advance understanding and knowledge in some concrete way. In part, the possible endless spinning of types is one reason some scholars such as Ralph Turner have taken a very negative view of the use of typologies and taxonomies in disaster research. As he has written:

> There are several reasons why I believe that preoccupation with the taxonomic exercise is likely to be counterproductive. First, sociology has a rich legacy of taxonomic systems that have been abandoned after initial enthusiasm gave way to disappointment over their unproductiveness. Second, sociological variables and types lack the critical characteristics that have made definitive taxonomies valuable in other fields of study. Third, useful classificatory schemes typically develop after theories are formulated or as they are formed rather than before. Fourth, precise classification is a critical tool in

> testing rather than in developing theory. Fifth, it is erroneous to expect one grand classificatory scheme to be equally useful in answering different questions. Sixth, theorizing that begins with classification usually ends with classification, producing static rather than dynamic end products. Finally, I believe that overriding search for an integrating disaster paradigm is based on confusion between fields of study that are defined by problem, process, or theory, and fields defined by substance.
>
> (1989: 266)

Certainly these are serious points that must be considered, especially his last one which goes to the heart of such an endeavor. (Although the issue is partly addressed by Stallings in his chapter, it would be worthwhile for some scholar to very systematically analyze the implications of the definitional objection raised by Turner for the disaster area.) Nevertheless, I would question the full applicability of all the objections to the disaster area. Also, the criticisms would particularly seem to be most applicable if the development of a typology was the only way being advocated for theoretical developments in the area. So without being dismissive of Turner's very sophisticated critique, it should not totally discourage those trying to make advances on the typological front.

Therefore, at this point, I will partly discuss what is involved in trying to develop a typology of crises (of which disasters are only one type). In short, I turn now to a discussion of the second of the two typologies I think are needed. Since I have discussed elsewhere (see, Quarantelli 1987b: 25–27) the value and kind of typologies needed about disaster phenomena themselves, I will here forego any further description and analysis.

Typologies can be of different kinds: empirical, logical, heuristic, ideal type, etc. My preference for disaster research is for the last kind. For those unaware of Weber's discussion of ideal types, he basically argues that one can develop types that project social phenomena as they would be if they existed in highly exaggerated or in pure form. As such as Weber himself has written:

> In its conceptual purity, this mental construct . . . cannot be found empirically anywhere in reality.
>
> (Weber 1947: 90)

However, such formulations have been used quite successfully in various areas of scientific inquiry. Actually a case can be made that many conceptions in disaster research, although never labeled ideal types as such, only make much sense if viewed as ideal types. In my view, for instance, the widely used four phases or stages of disasters, namely, mitigation, preparedness, response, and recovery, are best thought as ideal types because in terms of empirical reality, the phases have no clear-cut boundaries, often overlap, and depend for labeling upon which particular organizational entity is the focus of attention in any given study (Neal 1997). Similarly, I once advanced a fourfold typology of disaster sheltering and housing: emergency shelter, temporary shelter, temporary housing, and permanent housing (an early version is in Quarantelli 1985a; a later one in 1995d). Although used in empirical research (Bolin 1994), I would say that it too makes the greatest sense if thought of as an ideal type depiction. An initial typology of crises might best start with ideal types.

Ideal types can and should eventually be given empirical substance, and there can also be empirically derived types. The last approach in my view however is likely to open a Pandora's box of myriad types rather than contributing to understanding and knowledge, although it is my impression most disaster researchers would want to go the empirical route. However, this is not unrelated to how they conceptualize the central concept of disaster in the first place. So if one believes that elephants (disasters) are out there waiting to be examined, then the empirical route makes the most sense. If one is more inclined to think that disasters are socially constructed, that is not necessarily the case.

Now although I advocate the development of a typology of crises, I do not present one. I will not remotely attempt even the rough kind of categorization set forth by Barton. That would be a Herculean task. Instead, my intent is to ruminate about preliminary inquiries that I think should be made by anyone with a serious interest in developing a typology. Thus, there are no final answers or conclusions here, but primarily suggestions and hints on how someone theoretically inclined might proceed. In fact, there may not even be total consistency in what I say. This stems less from constant mentimutation, than from my not knowing what position might be most valid and useful for research purposes. Furthermore, my focus will primarily be on issues most relevant to the disaster subtype within the larger category of crises. Put another way, the conception of disaster used has to be in relative harmony with whatever is advanced as the larger concept of crisis.

Six typological issues that need addressing

As I see it, there are at least six complicated issues that have to be addressed, namely, whether:

1 To include sudden and/or chronic situations?
2 To use an agent or response focus?
3 Famines, epidemics, and droughts (FEDs), the older type of diffuse situations, should be placed within or outside the disaster label or even in the crisis typology?

4 The newer types of diffuse situations, such as computer system failures, should be categorized as disasters or not?
5 What currently are called *complex emergencies* should be in the typology at all?
6 Implicit value positions (for example, that disasters are *bad*) should be made explicit?

Inclusion of sudden and/or chronic situations?

A crucial question for the development of a crisis typology is whether to include both chronic and sudden situations. Assuming both are included in the larger framework, then there is the question of whether both should also be included within the subtype of disaster. Researchers have taken all possible positions on this matter, although some such as Kroll-Smith/Gunter seem to be implying that the distinction is an increasingly dubious one.

A number of different issues are involved in addressing this matter. Let me first suggest something that needs full exploration, and then indicate two lines of argument that could be pursued.

Every now and then I have criticized researchers for a failure to use *social* time rather than chronological time in their descriptions and analyses of disasters (most recently, Quarantelli 1994). In fact, to my knowledge, not one researcher has ever explicitly used social time in studying disasters, all the more a surprising failure on the part of those who operate with social constructionist views of concepts. The social aspects of time have been looked at for decades (see Sorokin 1943; Gurvitch 1964). Maybe the concept of social time is a complex one for research purposes, but at heart it has to do with how "time" is differentially experienced and visualized by individuals and groups (Flaherty 1993), as well as with the varying conceptions of temporal processes that are part of the cultural frameworks of different social systems (McGrath 1988). Passing references to "social time" by some of the authors in this volume come nowhere near this idea.

There is a substantial empirical and theoretical literature in psychology, sociology, and anthropology which documents that not only can a conceptual distinction be made between chronological and social time in studying any behavior, but that there is considerable value in so doing. Too much chronological time has now passed to further delay using the concept of social time in looking at disaster phenomena. Clearly, the application of the concept of social time would have direct relevance on whether and how researchers might approach the sudden-chronic issue. I do not claim it will be easy to do so, but there is a large body of existing knowledge ready for exploratory use and application (see, for example, Zerubavel 1981; McGrath and Kelly 1986; Young 1988; Young and Schuller 1988; Pronovost-Giles 1989; Baker 1993; Gosden 1994; Nowotny 1994; Adams 1995). If social rather than chronological times were to be incorporated into the theoretical view of disaster, the current separation made between sudden and chronic, drawn from everyday chronological notions, might recede into an insignificant difference for research purposes. However, it is crucial to note that since sudden social time may not correspond at all to sudden chronological time there would still be crisis related phenomena that would suddenly appear.

As to the two lines of argument, the first has to do with another large body of literature. The other again concerns the actual mixture of ideas that I have said researchers actually use in making categorizations (for a review of how scientific knowledge is socially constructed see Shapin 1995).

There is a substantial body of research that uses the concepts of stress and crisis (sometimes to refer to separate phenomena, sometimes to the same), which has fed into a large theoretical literature on both. The issue here is not whether what has been produced under such labels is either valid or important. By almost any criteria, they can be so characterized.

The particular labels are in themselves unimportant, but the use of one term rather than another is far beyond a matter of semantic preference. I have used both concepts in my own research. However, my view is that the labels as sensitizing concepts have different connotations. This can be illustrated by looking at the bodies of literature involved. They reflect a different imagery of the phenomena they study. For example, the literature on stress comes mostly out of a psychological and mental health background, primarily uses the individual as the basic unit of analysis, and implies a continuing condition. The literature on crisis has much more of a political science and exercise of power background, is more likely to use macro-level entities or collective ones as the major analytical unit, and implies a sudden state of affairs that requires a quick resolvement (see Hermann 1972). These different connotations of the two terms, at least in English, do suggest to me that it is wise not to intermingle the two terms. Some labels are more euonymnistic than others.

However, even more important than the connotational difference is that the labels can be used to denote something crucial for understanding social behavior. On the basis of the mixture of empirical facts, theoretical notions and logical analyses discussed earlier, let me advance the following thesis.

Few would dispute that risks/hazards of all kinds are myriad (although it is an open question as far as I am concerned whether they have actually increased or simply our perception of them, see Douglas and Wildavsky (1982) and Douglas (1992)). But while omnipresent, relatively few of them ever actualize, that is, come into being. A risk or hazard in other words is a potential, a possibility, not an actuality or a happening. As Beck has well stated:

> risks essentially express a *future* component . . . have something to do with anticipation, with destruction that has not yet happened.
> (1992: 33)

Of interest for this discussion are that of the relatively few risks/hazards which may be realized, even fewer require immediate action. There are hazardous situations where a delay in response of days or even years will not be significant. These typically are instances where there is considerable lead time before the risk will manifest itself in its most dangerous form. Examples of relevance to the issue being addressed here would be the slow chemical poisonings as might be occasioned by asbestos, radiation contamination by radon, climatological pollution through acid rain, some health epidemics, coastal erosion and land subsidence, and all but the very last stages of some famines and droughts. For such situations to be defined as crises is dependent on the exercise of political power and can become matters of political dispute, as seen in the current AIDS pandemic in certain African countries. In one sense, the examples given typically are of chronic hazards rather than sudden risks. Such hazards may make for stress, but not for crises.

A collective crisis can be conceptualized as having three interrelated features: (1) a threat of some kind, involving something that the group values; (2) when the occasion occurs it is relatively unexpected, being abrupt, at least in social time; and (3) the need to collectively react for otherwise the effects are seen as likely to be even more negative if nothing is done sooner or later (Quarantelli 1995a; see also Keown-McMullan 1997). That is, whatever the negative effects projected for the group, there will be quick, certain, and immediate economic, political, psychological, social, ecological, and symbolic costs if no collective remedial actions are attempted. As I have tried to state it here, an effort is made to avoid what Beck (1992: 100) says is the strong tendency in modern societies to individualize social risks.

Therefore, my view is that relatively sudden and relatively chronic risk/hazard actualizations ought to be distinguished from one another. I realize some see the difference between acute and chronic threats as disappearing (totally apart from using the notion of social time as I discussed earlier), but I am convinced that if the "sky is falling" in fifty years' time rather than tomorrow, even the strongest advocate of a disappearing distinction will not react the same to both projections. However, my major reasoning is that the conditions for, the careers of, and the consequences from the two kinds of situations are rather different from one another with the acute or sudden aspect being very vital in the social response.

To argue that for conceptual purposes we should look at sudden onset impacts (as I discuss later, actually to the responses to such occasions), in my view is not inconsistent with Rosenthal's argument that disasters cannot be understood unless a researcher looks far before and long after the happening. I agree strongly with his processual emphasis, a focus with which Perry, too, is impressed. A sudden manifestation does not mean that there cannot be a very long incubation period nor a long delayed effect. Vaughan's (1996) analysis of the *Challenger* explosion well illustrates the past roots of that disaster. To me the sudden manifestation is what mostly distinguished disasters from chronic hazards.

Thus, for substantive rather than semantic reasons my preference is for a typology of crises rather than a typology of stresses. If so, then disasters also have to be conceptualized as involved crises rather than stress. But regardless of the validity of my reasoning and conclusion, eventually researchers will have to decide whether they want to develop a typology that encompasses both crises and stresses or only one, and if one only, which one? My view is that as a start we should have a larger typology in terms of sudden rather than chronic situations, because that is the essence of a crisis (however that does not preclude having a larger typology of stress that has within it crises).

An agent or response focus?

Another important question in developing typologies is whether to include an agent or response focus. Unlike the sudden–chronic dimension that can be either/or, this has to be one or the other. Again, disaster researchers have taken both positions, although historically a case might be made for a slow shift from an agent to a response focus as Gilbert notes in his chapter.

In the early days of study there was a conflation of hazards and disasters. It could also be argued that the natural/technological disaster distinction is rooted in that starting point too. In both cases, the focus is on the supposed agent involved, a phenotypical approach. To be sure, many researchers continue to focus on agents. The many vested interest and professional groups that are specific agent focused insure that such an orientation will not soon disappear. But, as I see it, that is the past and will not be the future as social science researchers move more in a genotypical direction.

However, an abandonment of the natural/technological agent distinction does beg the question of where the focus of disasters ought to be sought. To some social scientists, an easy answer is to see disasters as being the social behaviors that appear in certain circumstances, independent of sometime prior physical happenings or conditions. This is another way of saying that a disaster is not a physical happening but social behavior that has certain characteristics. If so, we should avoid the very sloppy statements at times found in the literature along the line that a disaster led to or resulted in casualties or property destruction, as if there were a prior X (i.e., a disaster) which had certain effects. Apart from reifying an intellectual tool, a concept, might it not be better to visualize disasters as not having existence independent of the behavior of individuals and groups? As a colleague of mine once said, disasters "don't cause" anything. They are behaviors that come out of certain prior social conditions, a point that Dombrowsky also made in his chapter.

We should conceptualize disasters in exclusively social terms. There is strong resistance to accepting such a view, as witnessed by the fact that, for example, such researchers in this volume as Kreps, Stallings, and Oliver-Smith, who otherwise differ from one another along many lines, all explicitly maintain that we necessarily need to maintain a physical referent in defining disaster. Yet in certain of our approaches we already have divorced the physical from the social. For instance, when everyday emergencies are distinguished from disasters, as many if not most researchers do, the agent that may be involved is totally ignored and the characteristics of the response are emphasized (Dynes 1974, 1994a, 1994b). That is, the behavior rather than the agent is seen as the distinguishing feature around which to conceptualize disasters as something different from everyday emergencies.

That is good as far as I am concerned. The more we get away from our hang-up of including an agent or the physical environment as part of our conceptual view and focus on the social behavior involved, the better off we will be in at least five ways. First, it will undermine the frequent conflation of hazards with disasters. Second, it allows us to conceptualize as disasters significant crisis-type phenomena that have little or no connection to nature or the physical world. Third, it will allow us better to capture disasters of the future. Fourth, it will allow us to better distinguish *disasters* and *catastrophes*, phenomena which are behaviorally different especially in the response aspects (for some behavioral differences, see Quarantelli 1995a), but this is not the place to discuss the many values of such a conceptual differentiation. Fifth, it makes it easier to distinguish the conditions for something, what is produced from that something, and what follows from that production, from one another, a previous mantra of mine some might recognize.

I should also note that within that kind of framework there is an additional question of where the locus of disasters might be specifically placed. Dynes, for example, puts it at the community level for the most part. Perrow, although not focusing on disasters per se, makes a strong case that organizations are the building blocks of modern societies and therefore should be the prime focus of research activity on environmentally related stresses and crises, whether this be the pollution of Lake Baikal in the former Soviet Union or the ineffectual, tardy response to Chernobyl (1997: 67; see also Weick 1990, 1993; LaPorte and Consolini 1991). Put another way, a good conceptualization of disaster should specify the social unit within which the behavior is occurring, a not easy task as noted by Rosenthal in his chapter.

At another level, there is something very significant about the historical fact that while the human race generally has always attempted emergency management of crises, it evolved for eons without undertaking much if any large-scale and continuous disaster planning. Yet humans have not only survived, but, by most criteria that could be applied, have increasingly prospered. So what is it about the last century which has led to the increasing emergence of disaster planning? Along some lines, I think this is also telling us something about how we go about thinking about disasters. It is, for one, linked to the increasing complexity of social organization, which of course leads us necessarily in the behavioral rather than agent direction.

Clearly what positions are taken on the agent–response dimension just discussed will have important implications for developing both a typology of crises and of disasters. In my view, a response focus is where we are heading, so the direction is right. However, the pace should be quickened.

The placement of older type diffuse happenings?

Since the dawn of human history, observers of the social scene have often noted the recurrent appearance of certain, diffused in space and time, negatively viewed happenings to which such labels as famines, epidemics, and droughts (FEDs) have been applied. In popular parlance, and even more so in the more technical/professional literature, these have frequently been labeled "disasters."

However, there are some puzzling reactions by self-styled disaster researchers to these kinds of phenomena as old as human history. On the one hand, most seem very reluctant, for example, not to call the famine currently occurring in Sudan a "disaster." In fact, at least in passing, FEDs are often listed as part of what the field should study. That the researchers almost never do such studies may be because the great majority are from developed rather than developing societies where FEDs are far more common (exceptions are mostly applied disaster researchers in England primarily interested in international relief, and who often publish in the journal *Disasters*).

On the other hand, disaster researchers have maintained a remarkable professional distance from the workers and the field that directly focus on and study FEDs. There is a vast literature on FEDs, probably more in totality than in the disaster literature per se. As seen in this volume, with only isolated exceptions (Oliver-Smith being the major one), the publications are ignored and almost never referenced. In addition, the past/present leading FED scholars are not part of the social circles of disaster researchers. As a simple example of the latter, there is a research committee with the name of The Famine and Society Thematic Group that has almost no overlap in membership with the Research Committee on Disaster in the International Sociological Association of which both are a part (at one time I was the only researcher who concurrently was a member of both groups!).

Whatever the past and current treatment by researchers, FEDs need to be categorized, both as to whether they can be meaningfully placed under the disaster rubric or if they even belong in a crisis typology at all. At present, my general inclination would be to exclude FEDs from the disaster category and to treat FEDs as social problems, involving chronic stress settings rather than crisis occasions. This I say for the following two reasons.

First, FEDs are diffuse happenings both in terms of chronological time and geographic space. Just at that level they differ from relatively focused time and space–time limited occasions. They also lack the suddenness discussed earlier. Equally as important, most FEDs can only be identified in terms of response since usually (except in some epidemics such as in the Ebola Virus infections or locust infestations) there are no agents involved in any way comparable to an earthquake or a gas explosion. As behaviors, FEDs seem to me to differ even phenotypically from sudden disasters.

Second, and supportive of what has just been said, many well-established observations findings about disaster behavior at any level of analysis, simply do not appear to apply to these diffuse kinds of happenings. There is a vast literature on famine and droughts (see recently, for example, Harrison 1988; Devereux 1993; Nikiforuk 1993; Kohn 1995; and for one of the earliest disaster scholars who looked at famines, see Sorokin 1942). Looking at the FEDs literature and applying it to sudden disasters, as I have very informally and unsystematically done, shows that the empirical data and theoretical ideas do not overlap very much (see, for example, Glantz 1976; Walker 1989; Downs, Kerner and Reyna 1991; Deng and Minear 1992; Field 1993). In this respect, McCann's (1987) discussion of the vulnerability of northeastern Ethiopia to famine despite it having one of Africa's most efficient traditional agricultural systems impressed me very much. However, beyond historical case studies, a systematic point-by-point comparison of research findings from the FEDs and the disaster literature should be done. A start could be made by looking at so-called warnings for disasters and famines (the Buchanan-Smith and Davies (1995) book would provide a good starting point for such a comparison).

From my perspective, a good case can be made that the stressful kinds of FEDs discussed should not be conceptualized as disasters, and probably not even as instances of crises (although there may be instances here of what I mentioned earlier, namely that sometimes genotypical characterizations may cut across phenotypical appearances, as could be true in the instance of some famines).

The placement of newer type future crises?

I have long argued that we should in refining our conceptual tools look at likely future disastrous occasions rather than looking back only at past disasters. A basic finding of researchers about preparedness is that unfortunately too many communities take the last disaster that has happened in their area as the prototype around which to plan, instead of projecting for planning purposes the more likely kind of occasion that will occur in the future. Should we also, as disaster researchers, not project more into the future rather than looking at just the past and the present? Projecting into the future suggests that human societies will probably be faced with more and worse crises as a result of the worldwide maturation of industrialization and urbanization processes (see Quarantelli 1996). In particular, there will be instances where localities will have disastrous conditions from quite distant sources, and there also can be a catastrophic potential even if there are no casualties or major physical destruction. However, contrary to some speculations, not all of the future threats will be invisible and nonperceptible. In any case, the features indicated ought to affect response behavior.

Past, present, and future examples can be cited. We already have had radiation fallout in countries far distant from Chernobyl, as well as the contamination in the Rhine River that affected five nations. The world has already seen computer system failures and related problems such as the Barings Bank collapse in England (Fay 1996) or the copper trade losses in Japan that amounted to 1.8 billion dollars. With cloning now being done for medical and animal husbandry purposes, as I noted more than a decade ago, it is a question only of when and not if we will have disasters of a biogenetic nature (and in the late twenty-first century even more dramatic neurotechnological ones). The kinds of crises involved all have had or could have very major and widespread negative economic, social, psychological, political, ecological, and symbolic effects with few if any casualties and/or little major property damage. The effects frequently go far beyond the site of the happening and may be national or international in scope. This, we have already seen in occasions ranging from a suddenly bankrupt Orange County, California to the explosion of the space shuttle *Challenger* (Presidential Commission 1986). The financial kinds of crises have also started to catch the attention of government agencies as witnessed in a recent report that used research findings from the disaster area to partly analyze the Mexico debt crisis of 1982, the Continental Illinois bank crisis, the Ohio savings and loan crisis and the Stock Market Crisis of 1987 (see United States General Accounting Office 1997).

One interesting implication of this is that perhaps such future occasions might impact more on the better off in social systems, contrary to what most research presently indicates. (However, some past studies have already shown, for example, that it was the middle socioeconomic levels that were more impacted than those in lower status levels in the 1985 Mexico City earthquake, as documented in Dynes, Quarantelli and Wenger (1990). Also, clearly some recurrent brush fires in recent decades in certain suburban areas of Los Angeles and around the Riviera in France have primarily impacted on high status households.) Nevertheless, I raise this point because if we define or conceptualize disasters in terms of some aspects of the social response, should the social status of those most affected matter or not matter in that conceptualization? As such, we should think beyond individual victims. It seems clear to me that the collective reactions to the *Challenger* explosion were determined more by the social status of the social system and the institutional sectors and the organizations involved, than by the fact that there were seven dead crew members.

Actually, I have no problem in thinking that we should conceptualize these newer kinds of crises as disasters. That is not the major conceptual problem I see. Rather, there is a question in my mind about whether the responses to these newer crises have enough characteristics that might be viewed as distinctive (e.g., reactions to distant sources, non-deadly catastrophic consequences, macro-level organizational involvement) so that we may have to treat them as a singular subtype within the larger type of disaster within a crisis typology. Others will have to deal with this very interesting issue that I also think will make a genotypical conceptualization and the use of social space–time conceptions even more necessary than at present.

The placement of complex emergencies*?*

One question that those interested in typologizing disasters and crises need to deal with has primarily been generated outside the research area. The reference here is to what are often called in the terminology of the United Nations *complex emergencies* (or sometimes *compound disasters*). These labels have been increasingly used by those who are primarily involved in the provision of international relief and humanitarian assistance to victims of widespread and multi-related phenomena that result in extensive human suffering and misery (see Moore 1996). Good current examples would be Afghanistan and Rwanda (and earlier Cambodia and Mozambique) where simultaneously there are complex mixtures of civil strife, famines, genocidal activities, epidemics, and large-scale displacement and movement of refugees.

Where are such happenings to be classified in typologies of crises and disasters? From my research perspective on disasters, they could be dismissed as being primarily conflict situations (which however does not solve the problem for those categorizing conflicts as disasters). Certainly violent struggle is always involved in the examples given and therefore, if my previous discussion of conflicts is accepted, would have to be the concern at best of those attempting to develop crisis typologies.

However, I am not inclined to dismiss totally these happenings. They can be used to raise a broader issue, that is, the larger social context within which disasters happen. Disaster research so far is clearly underemphasizing the larger social setting. This became clear to me at a conference of former Soviet Union and American disaster researchers in Moscow a few years ago One of the Russians, after hearing about research results from American society concerning who and what was done in the disaster recovery phase, expressed doubt that the same results would be found in his country. His explanation, which rang true, was that the studies had not been done in societies with everyday stressful turbulent social settings such as Russia (Kartashov 1991). In his view, a disaster that might be deemed major in the West could be a rather minor occasion compared with other crises in the system. My perception of the importance of the larger context was reinforced when recently I helped Porfiriev assemble a collection of current disaster studies in the former Soviet Union (Porfiriev and Quarantelli 1996). It is very clear from the dozen reports in that volume that the larger social context fundamentally affects very much the views of Russian researchers about the nature of disasters.

Actually, it is not necessary to go that far in terms of different kinds of social systems. As Ino Rossi documents, the reconstruction after the 1993 earthquake in southern Italy can only be explained by using the larger context of Italian society, its structure and values (including family and kinship linkages, political realities, acceptable levels of corruption), rather than primarily by whatever was the immediate social response to extensive physical damage and destruction (Rossi 1994). Here, too, the negative costs or effects in many ways to communities and the society in the recovery phase was far higher than the disaster itself (which is similar to some of what Rosenthal discusses in his chapter).

So, whether called *complex emergencies* or something else, I do think disaster researchers should systematically look at these kinds of situations because they necessarily call attention to the larger social context or setting of disasters. I am not aware of even one who has done so. We would do a better job of typologizing disasters and crises if we did so.

Let me build on this point to make a related but much larger one. Researchers in the area have tended to ignore the larger social context or macro-level social setting of disasters. Currently, a massive transformation of social life is occurring (see Smelser 1991; Tiryakian 1994). Enormous social changes are happening in the political, economic, familial, cultural, educational, and scientific areas everywhere in this world. As examples, we can note the new family and household patterns that are emerging, the basic alterations occurring in the role and status of women, the move almost everywhere to a market-type economy to produce goods and distribute services, the spread of at least nominal democratic patterns of government, the growing dominance of nontraditional artistic and musical forms as well as globalization of popular culture, the escalating employment of computers and related means for training and educating people, and the growing diffusion and expanding use of applied social science to many areas of life. Among other things, clearly these changes will transform the number and kinds of disasters that will occur and the nature of the disaster planning and managing necessary in the future.

Now, it is true that a few of the authors in this volume allude to some such changes, but as a whole the major social transformations occurring have not been systematically incorporated into theoretical ideas about disasters. We in the field need to pay much more explicit attention to the larger social context of disasters. Dealing with the related crises of *complex emergencies* might give us some empirically grounded clues on how to go about incorporating relevant ideas into our conceptual views of disasters.

Implicit or explicit value positions?

Some traditionally held assumptions about the supposed characteristics of disasters have always puzzled me, although I have not always fully articulated them even to myself. However, one that I came to make more explicit recently, and partly because of working on this volume, is that certain value positions regarding most crises and all disasters are almost never, if at all, examined. There is, for example, the assumption (if not actually part of the definition, such as in Saylor 1993) that disasters have negative outcomes associated with them. Disaster researchers generally have had as part of their paradigm the notion that disasters whatever they are, are *bad*. That clearly implies some value position.

Taking such a position seems to stand out more clearly with regard to conflict situations. For example, are the social effects of the Russian military intervention in Chechnya classifiable as "bad"? What about what happened in Tiananmen Square in Beijing, China? Who was it "bad" for? Was it "good" for anyone? What criteria are being used? What of the ongoing Turkish attack on the Kurds in northern Iraq or conversely the rebellion of the Kurds against the present Iraqi regime? What about the hostage-taking in Peru? My guess is that the answer in many cases would depend on the political sympathies of those making the classification. As a long time student of collective behavior and social movements, I have noted even among social scientists that the characterization of certain happenings as *riots* or *rebellions*, or as *terrorist attacks* or *liberation movements* have sometimes reflected not disciplinary tools and concepts, but extra scientific consideration. Goodness or badness is in the eyes of the beholders, not somehow out there in the world.

Why do we not see this in the instance of disasters and crises? Why are they "bad"? Only a very few have even slightly challenged this badness assumption, such as Scanlon (1988) who empirically shows that there are "winners" as well as "losers" associated with disasters; see also Dahlhamer and Tierney (1996) and also some first class reports by newspaper reporters such as Feder (1993). Interestingly, there are economic analyses that at least have questioned whether on balance there are always more economic losses than gains, perhaps reflecting a statement by John Stuart Mill more than a century ago that disasters merely consumed over a short period of time what would in the long run nonetheless always be used up (but for a strong counter argument, see Hazards Assessment Update (1996), also Cochrane (1975)). Even the "benefits" of famine have been discussed (see Keen 1994). Certainly, anyone with a sociological background ought to understand that any social phenomena will have functional *and* dysfunctional consequences.

My view is that whatever the use of a conception of "badness" for operational relief agencies, it may not be the most strategic conceptual position that researchers should take. It seems to me that the negative consequences of disasters should be a matter of empirical findings and not an a priori part of a definitional concept (which is a common-sense notion). Not only will this allow us to see better the positive social aspects of disasters, it will also permit us to treat threshold problems, if any, as both consequences and conditions for disaster phenomena. In addition, it would allow researchers to move further away from physical aspects and move even more toward clear-cut social dimensions.

Let me conclude this section of the chapter by saying I have tried to build on what I understood was said by the authors in this volume and in the past literature. Suggestions have been made where we, as a field, might go. Perhaps the paths recommended might to some seem mixed and possibly not always consistent with one another. Although not intended, that could be the case because much of what I suggest should be done (e.g. using social space–time dimensions, genotypical features, etc.) is not currently in place. However, while I may be not altogether certain of the answers suggested or implied, I do feel very strongly that the questions I have raised must be dealt with in some way or other by anyone interested in the conceptualization of what is a disaster. At one level what has been offered could make a significant difference, although not as much as what I will now suggest in the following section.

A NEW PARADIGM?

Do we need a new paradigm for the field of disaster research? In dealing with this question let me first note that I do not see many researchers either indicating a need or pressing for the development of new paradigms. A few authors in this volume and others elsewhere might dispute this. However, to me the major tendency has been, and is, to tinker with ideas traditionally used by researchers. They are, to use a typology from the field of collective behavior, reformers rather than revolutionaries.

I say this while recognizing that scholars in the sociology of science would probably argue that the field of disaster studies has already undergone a paradigmatic shift (Gilbert, in his chapter, also seems to claim that historically there have been three paradigmatic shifts in the area). The earliest workers in the area, including myself, with little conscious thought and accepting common sense views, initially accepted as a prototype model the notion that disasters were an outside attack upon social systems that "broke down" in the face of such an assault from outside. Many of us doing the pioneer research very quickly moved to a different model in which disasters were generated from within social systems themselves and the response to them had an emergent (building-up) quality. Maybe empirical data forced the shift. However, in my case it was more a result of subtle theoretical criticisms from insightful European researchers such as Carlo Pelanda and Wolf Dombrowsky, and also the fact that a very similar shift from a "breakdown" to an internal emergent structural shift occurred, starting in the 1960s in the sociological subfield of collective behavior and social movements.

Whatever accounted for it, in my view, there was a paradigmatic shift. A Ptolemaic-like conception of disasters was replaced by a more Copernican one. There was a shift in locus–focus from the physical environment (victims seeing or having hazards impacting them) to one making the social system central to any analysis. Actually, to this day, many *hazard* researchers in contrast to most *disaster* researchers implicitly still use the old paradigm. Also, even if they do not seem to recognize it, those who see technological disasters as distinctive also lean that way, as well as some but not all with postmodernistic views. Apparently "victims" are still the center of the disaster universe for some students in the area. However, even those who use some postmodernistic notions, which are seen at variance with traditional views of disasters, are not in my judgment that radical (for a totally contrary view about postmodernism, see Owen (1997) who discusses how postmodernistic ideas are forcing sociology to rethink its central conventional categories and practices).

This is not the place for me to indicate my views about the claims of postmodernism with regard to the social sciences (see, for example, Doherty, Graham, and Malek 1992; Dickens and Fontana 1994) after its partial initial emergence from artistic and humanistic fields of study. Others more interested and knowledgeable regarding that critique have well expressed some philosophical and practical problems with that particular view of the world, especially its nihilistic particularism. However, some postmodernistic notions have been applied in the disaster area. For the most part, those applications have left me unimpressed even when they have gone beyond just objecting to whatever is done in the name of a scientific framework and projecting dystopias. I say this because to me those ideas are not much more than an oversimplistic extension to an extreme of basic ideas and notions already deeply embedded in Western philosophy and science.

For example, there is the idea of looking at social phenomena from the viewpoint of the actor, in the case of disasters, of victims. However, taking into account the subjective perspective of social actors is a question much discussed in ancient Greek thought. For scholars in sociology, an interpretative understanding of behavior is a very old idea, expressed by Max Weber among others, although perhaps new in the emergence of sociology as a discipline. Furthermore, as someone trained as a social psychologist in a Meadian symbolic interactionist framework, I see using the actor's perspective to understand social phenomena as a very old and traditional view, and useful in its place, even in disaster research (such as my use of this perspective in understanding panic flight behavior, see Quarantelli 1954, 1957). (In retrospect, I see that one very justifiable criticism of my work on panic would be that I mostly developed a social psychological analysis of panic flight from the viewpoint of the participants. I failed to develop a more sociological one in terms of group disruption, both in terms of social relationships and social climate when these are thought of in holistic rather than reductionistic terms, but which are well examined in later work on panic by Johnson (1988), Johnston and Johnson (1989) and Johnson, Feinberg, and Johnston (1994)). As such, my view is that postmodernism in some of its major tenets, at least in their application to the disaster area, seem to be simply an overextension of ideas already lurking in traditional Western thought.

What if we wanted to go beyond such traditional views? Perhaps one good way to start moving in such a direction is to look at the ideas and perspectives and implications of adopting some currently much talked about intellectual trends in the social sciences, particularly if they seem relevant to disaster topics. All would seem to require a paradigmatic shift. Examples might be:

1 *Chaos theory* (see Gleick 1987; Waldrop 1992; Gregersen and Sailer 1993; Guastello 1995; Robinson and Combs 1995; and Kiel and Elliott 1996; for an attempt to apply it to disaster phenomena, see Koehler (1995)). My general impression is that this might be a particularly good candidate for extensive examination in the near future, although *not* at all because of the homologous label used, but because it addresses fundamental problems of uncertainty and ambiguity in social settings that are features of the emergency time periods of disasters. As also asked of social phenomena generally (see Baldus 1990), whether the probable nonlinearity of processes in disasters makes them more difficult to fully study in a positivistic framework, would seem an open question. Along another line, Smith (1991) raises the fascinating question as to whether chaos theory might allow better social science research but make social engineering a much more dubious enterprise. Applied to the disaster area that certainly would be a radically different perspective on social phenomena such as disasters.
2 *Ecofeminist views* (especially the deep and radical ecology writings and others which discuss human relationships to the environment, see Plumwood (1993), Warren (1994), Zimmerman (1994), Forbes and Sells (1996), Merchant (1996), Rocheleau, Thomas-Slayter, and Wangari (1996)). Both from a theoretical and methodological viewpoint, feminist scholarship has sharply questioned traditional views of and approaches to social phenomena. If nothing else, this theoretical view has forced researchers to think about traditional assumptions, ask previously not thought about questions, and suggests new perspectives on whole areas of social life. As such, this could surface another way of looking at disaster phenomena. I am not aware of the application of this

view on clear-cut disaster phenomena, but it has been applied to the "mad cow" disease (Adams 1997) and to technological risk perceptions (Cutter, Tiefenbacher, and Solecki 1992). The concept of gender that some disaster researchers have started to use (see Fothergill (1996) for a review of such studies; see also Anderson and Manuel 1994; Morrow and Enarson 1996; Phillips 1990; Scanlon 1997) while definitively a worthwhile move, is only a very small step in the possible application of the more radical ecofeminist scholarship.

3 *Cultural theories*. The concept of culture is a central one in anthropology. Its application to the disaster area is well illustrated in Oliver-Smiths' chapter in this volume; see also his review article (1996) where other anthropological applications of cultural ideas to crises are noted. However, it is only recently that some of the relevant ideas about culture have spread very widely in the other social sciences (see, for example, Featherstone (1995), Milton (1996), and Fowler (1997) who builds on Bourdieu), although little has been applied so far to any kind of crisis behavior (although it has been used by Douglas and Wildavsky (1982) to analyze how risk is viewed). Notions of culture and related conceptions have recently spread widely in sociology and on such topics as social movements bringing it closer to disaster phenomena (e.g., Johnston and Klandermans 1995). It might also be interesting to explore for disaster research the implications of formulations such as set forth by Thompson, Ellis, and Wildavsky (1990) that typologize ways of life into egalitarianism, fatalism, individualism, hierarchy, and autonomy. If the behavioral response is the crux of a disaster, then it would be certainly influenced by the cultural ways of living embedded in the situation. Furthermore, notions of culture could perhaps act as a supplement to more structural approaches and help undermine the functional bias sometime attributed to disaster researchers (but not all such researchers seem to be fully aware of recent theoretical advances in social structure analysis they could use, such as summarized in Prendergast and Knottnerous (1994)).

4 *Theoretical approaches to risk*. As I use this label, it covers heterogeneous clusters of only partly overlapping views whose most common element is a concern with the dangers and threats that face the modern world with its strong technological base, and often also with the idea that citizens and experts disagree on risks (for the latter, see Margolis 1996). The views include the very abstract work of Beck (1992, 1995, 1997) and of Luhmann (1993) who differ from one another, to statements such as by Eder (1996) who critiques both Marx and Durkheim, as well as counter-reactions to the major Beck thesis (e.g., Mol and Spaargaren 1993; Cohen 1996) that the relationship between individuals and hazardous technology is coming to replace conventional definitions of social class as the primary determinant of social stratification. At present

the approaches tend to focus almost exclusively on risks associated with technology in post industrial societies and lack much empirical grounding (see Tierney, forthcoming). Also, as I see it, some of these views have been applied to hazards but, as also implied by Stallings earlier, only occasionally to disasters as such. Nevertheless, a promising aspect is that much is visualized in the framework of social change (for example, Beck advances a macrosociology of social change), and there is an attempt to go beyond postmodernism with such concepts as reflexive modernization and autopoietic subsystems (i.e., the notion that society is composed of subsystems that behave according to their own internal codes). Clearly, there are ideas in this general approach that could generate new conceptions of disasters.

5 *Theoretical formulations that stress emotional–affective factors*. Emotions, with a sometime exception of psychology, have always had a marginal conceptual position in the social sciences (Lewis and Haviland 1993). However, some recent work has advanced some innovative ideas (see Mestrovic (1996), who discusses the post-emotional society; see also Harre 1986; Franks and McCarthy 1989; Lewis and Haviland 1993; Ekman and Davidson 1994; Kitayama and Markus 1994). Now if disasters/crises involve affect, which of course they do, perhaps a look at such a perspective might be useful (alluded to in passing in the chapter by Kroll-Smith/Gunter). Disaster researchers probably have de-emphasized feelings too much in favor of a cognitive stance. I can say this as someone who almost lost the fear dimension in my discussion of panic flight behavior because of an interest in trying to emphasize that all behavior is organized and always follows norms, as well as the usefulness of taking an actor's perspective into account in order to understand even seemingly "irrational" behavior.

To be sure, some of the five approaches noted above are nothing more than an ideological advocacy of certain semi-religious, philosophical, and political values and, to the chosen, revelations that they have adopted in an absolutistic way. Now advocacy of any kind of ideology does have its place; some societies and people have gained from ideological fanaticism, although there is often a negative downside as can be seen in the world today. Far more important is that the above approaches, in various ways, do raise very significant questions about the study of any kind of social phenomena. They should not be dismissed out of hand, which some operating in more traditional frameworks tend to do. Having openness of dialogue does not mean conversion. Disaster research especially ought to be open to examining the implications of any nontraditional view, particularly if it suggests a new paradigm. Of all people, researchers who recognize the importance and necessity of new and emergent phenomena at the crisis time of disasters, should be open to the emergence of nontraditional views!

Actually, we should go far beyond Western thought frameworks (represented in the approaches just listed). In looking for authors to write chapters for this volume, I sought, but failed, to find anyone working in the disaster area who used primarily a non-Western frame. Perhaps in a future addressing of the question of what is a disaster, such a quest ought to be more vigorously pursued.

It does not take much imagination to see that those cultures in certain Asian and African countries, with an emphasis on the collectivity or the family and extended kin group as being far more important than the individual, might very well come up with rather different conceptions of disaster. Similarly, cultures that emphasize that there are other valid ways of obtaining knowledge beyond the traditional scientific way, might also conceptualize disasters in ways quite at variance with traditional Western views. One does not have to agree with any of the most extreme postmodernistic critiques of science or even be a cult follower of the *Star Trek* science fiction series, to question if the currently accepted logical positivism structure of scientific research is the ultimate and only possible point of view regarding epistemology or ontology. Is the Western-derived scientific model the only social phenomenon in the world that is not subject to an alternative view or to change? Likewise, those cultural frameworks that take seriously the idea of cultural relativity, might also be able to visualize disasters in rather different ways than prevail in the West. Probably all of us are aware that some non-Western scholars in recent years using that very idea of cultural relativity (ironically a Western-derived notion), have increasingly challenged prevailing Western notions of journalism and press freedom, political democracy, appropriate gender roles and related behavior, health and medical treatment, artistic and popular culture expressions, and universal human values.

Perhaps it is time that the concept of disaster be subjected to the same kind of drastic scrutiny. The point, of course, is not to attack Western ideas in the politically correct way currently fashionable in some circles, but to see if for the purposes of disaster research, it might be possible to develop a new paradigm. Of course a new conception of disaster is only part of what might be involved in any paradigmatic shift of the field. Nevertheless, it could be a prime mover. As Huff has written:

> whether it be the discovery of oxygen, the perception of a new planet, or the positing of such constructs as the positron, the meson, or the neutrino, the history of natural science . . . repeatedly shows the central role played by concept formation. From this perspective, theoretical innovation is heavily indebted to the postulating or "conjecturing" of novel relationships between "old facts" and new entities: stated differently, innovation is the result of discovering

> new ways to conceptually organize previously known but puzzling and inexplicable phenomena.
>
> (1973: 261)

On the other hand, in developing new conceptions we might end back roughly where we started. In another paper (Quarantelli 1987b), I noted that many of us now blame "society" for disasters instead of seeing them as "acts of God." Yet any sociologist familiar with Durkheim will recognize that in substituting one label for another, that change in attributed source is perhaps less significant than may appear at first glance. For those unaware of his sociological analysis of religion, let it be simply said that he very imaginatively shows an equation of the human creation of the concept of God with the pressures of society on individuals (for a recent edition, see Durkheim 1995).

There is also the danger that in going for the new, there will be a break in continuity with the knowledge and understanding that the old has already provided. Some views expressed in this volume come perilously close to advocating such a break. But as Beck has written:

> we must retain good relations with the treasures of tradition, without a misconceived and sorrowful turn to the new, which always remains old anyway.
>
> (1992: 12)

Put another way, any new paradigm, if it is that different from what it replaces, will have problems of maintaining the historical and intellectual continuity that allows intelligent discourse to go on.

On the other hand, the more revolutionary we are in our thinking, the more likely we are to generate a new paradigm for disaster research. At least some of us ought to be revolutionaries rather than reformers. As a long time student of collective behavior and social movements, I am very well aware that the overwhelming majority of revolutions end in failure. But now and then one succeeds and transforms the behavior in the societies in which they occur, often in unexpected ways. So the more venturesome and imaginative among us should be encouraged to see if they can develop different paradigms for disaster research. If any such effort is successful (and probably even if not successful, since the counter-reactions could prove fruitful in advancing the dialogue), a future volume on this theoretical aspect of disasters might be markedly different from what this volume has covered.

Finally, I should note that there are several other issues, some raised by different authors in this volume, which also need to be addressed in the future. Three of them in particular strike me as needing priority of attention. The role of the mass media in "defining" disasters was highlighted, for example, in the chapter by Rosenthal, and also noted by others. Also, the importance of the political arena in influencing definitions of disaster was alluded to in different ways by several authors (which reminds me that unfortunately no one has ever followed up on the provocative semi-Marxist approach to disasters advocated a long time ago, namely in a collective behavior text by Brown and Goldin (1973)). Any theoretical analysis of "disaster" would seem to have to take into account how both mass communication systems and political systems strongly influence popular conceptions of the phenomena, as well as how disaster planners and managers think of disastrous occasions. Finally, given the implied and explicit statements of a number of authors in this volume, there is an interesting question about the degree to which any intellectual orthodoxy can coexist with multiple views about disasters. Simply as an example, can any extreme postmodernistic view really argue for the tolerance of many perspectives when it insists on a dogmatic acceptance of an underlying master orthodoxy that it assumes is the only basic way in which the world can be viewed? Or, as another example, can a scientific perception of reality allow for any non-positivistic views about disasters? There should be more systematic examination of such questions since, given what has been written in this volume, there appears to be no consensus on this matter also.

A LAST WORD

I started this chapter by noting that disaster researchers are operating with different assumptions about the nature of reality and knowledge. As such, they are bound to come up with and use different conceptions of disasters. Probably most readers will have noted that much of what was discussed in this chapter often implicitly rested on these fundamental differences in basic starting point. I am not sure that such implicit bedrock differences can be reconciled. In contrast, at least in principle it seems to me that many, although not all, of the explicit questions and issues in this chapter can be discussed and relative agreement or consensus reached. But axiomatic disagreements would seem to be of a different order. That is one reason why, although I raised these philosophical matters, I did not address them directly. We may all have to live with them. At a personal level this is certainly possible. I was trained as an elephant hunter. Because of being infected during my professional training with the Thomas virus of "if people define a situation as real, it is real insofar as consequences are concerned," I also came to believe in being a social constructionist. And the more I have lived, the more I have wondered that my unique insights about the world are not as equally obvious to others, including what I have written in this chapter.

BIBLIOGRAPHY

Adams, B. (1995) *Timewatch: The Social Analysis of Time*, Cambridge MA: Blackwell.

Adams, C. (1997) " 'Mad cow' disease and the animal industrial complex," *Organization and Environment* 10: 26–51.

Agency for International Development (1970) *Peru Earthquake: May 31, 1970*, Washington DC: Agency for International Development.

Alexander, D. (1993) *Natural Disasters*, New York: Chapman and Hall.

—— (1995) "A survey of the field of natural hazards and disaster studies," in A. Carrara and F. Guzzetti (eds) *Geographical Information Systems in Assessing Natural Hazards*, Boston: Kluwer Academic Publishers.

American Sociological Association (1996) *Preliminary Program*, Washington DC: American Sociological Association.

Anderson, K., and Manuel, G. (1994) "Gender differences in reported stress response to the Loma Prieta earthquake," *Sex Roles* 30: 9–10.

Anderson, K., Armitage, D. and Wittner, J. (1987) "Beginning where we: Feminist methodology in oral history," *Oral History Review* 15: 103–127.

Anderson, W. (1979) "Social science disaster research in the United States," *Emergency Planning Digest* (Jan/March): 20–24.

Arney, W. (1991) *Experts in the Age of Systems*, Albuquerque: University of New Mexico Press.

Aronoff, M. and Gunter, V. (1992) "Defining disaster: local constructions for recovery in the aftermath of chemical contamination," *Social Problems* 39: 345–365.

Babbie, E. (1973) *The Practice of Social Research*, Belmont CA: Wadsworth Publishing.

—— (1995) *Survey Research Methods*, Belmont CA: Wadsworth Publishing.

Bahtia, B. (1991) *Famines in India*, Delhi, India: Konark Publishers.

Bailey, K. (1989) "Taxonomy and disaster: prospects and problems," *International Journal of Mass Emergencies and Disasters* 7: 419–431.

—— (1992) "Typologies," in E. Borgatta and M. Borgatta (eds) *Encyclopedia of Sociology* New York: Macmillan.

Baker, P. (1993) "Space, time, space-time and society (space-time in the context of sociological and anthropological notions of space and time)," *Sociological Inquiry* 63: 406–424.

Baker, R. (1974) "Famine: the cost of development," *Ecologist* 4: 170–175.

Baldus, B. (1990) "Positivism's twilight?" *Canadian Journal of Sociology* 15: 149–159.

Ball, N. (1979) "Some notes on defining disaster: suggestion for a disaster continuum," *Disasters* 3: 3–7.

Bardo, J. (1978) "Organizational response to disaster: a typology of adaptation and change," *Mass Emergencies* 3: 87–104.

Barkun, M. (1974) *Disaster and the Millennium*, New Haven: Yale University Press.

Barton, A. (1963) *Social Organization Under Stress*, Washington DC: National Academy of Sciences.

—— (1969) *Communities in Disasters: A Sociological Analysis of Collective Stress Situations*, New York: Doubleday.

—— (1989) "Taxonomies of disaster and macrosocial theory," in G. Kreps (ed.) *Social Structure and Disaster*, Newark DE: University of Delaware Press.

Bates, F. and Peacock, W. (1987) "Disasters and social change," in R. Dynes, B. De Marchi, and C. Pelanda (eds) *Sociology of Disasters*, Milan, Italy: Franco Angeli.

—— (1989) "Long term recovery," *International Journal of Mass Emergencies and Disaster* 7: 349–365.

—— (1993) *Living Conditions, Disasters and Development: An Approach to Cross-Cultural Comparisons*, Athens GA: University of Georgia Press.

Bates, F. and Pelanda, C. (1994) "An ecological approach to disasters," in R. Dynes and K. Tierney (eds) *Disasters, Collective Behavior and Social Organization*, Newark DE: University of Delaware Press.

Bauman, Z. (1987) *Legislators and Interpreters*, Ithaca NY: Cornell University Press.

—— (1989) *Modernity and the Holocaust*, Ithaca NY: Cornell University Press.

—— (1993) *Postmodern Ethics*, Oxford: Blackwell.

Beck, U. (1992) *Risk Society: Towards A New Modernity*, Thousand Oaks CA: Sage.

—— (1995) *Ecological Politics in an Age of Risk*, Cambridge: Polity.

—— (1997) "Subpolitics: ecology and the disintegration of institutional power," *Organization and Environment* 10: 52–65.

Benesch, W. (1997) *An Introduction to Comparative Philosophy*, New York: St. Martin's Press.

Benini, A. (1993) "Simulation of the effectiveness of protection and assistance for victims of armed conflict (Sepavac): an example from Mali, West Africa," *Journal of Contingencies and Crisis Management* 1: 215–228.

Berger, J., Wagner, D., and Zelditch, M. (1989) "Theory growth, social processes and metatheory," in J. Turner (ed.) *Theory Building in Sociology*, Beverly Hills CA: Sage.

Berren, M., Beigel, A., and Ghertner, S. (1980) "A typology for the classification of disasters," *Community Mental Health Journal*, 16: 103–111.

Blaikie, P., Cannon, T., Davis, I., and Wisner, B. (1994) *At Risk: Natural Hazards, People's Vulnerability and Disasters*, London: Routledge.

Blalock, H. (1969) *Theory Construction*, Englewood Cliffs NJ: Prentice Hall.

—— (1982) *Conceptualization and Measurement in the Social Sciences*, Beverly Hills CA: Sage.

—— (1984) *Basic Dilemmas in the Social Sciences*, Beverly Hills CA: Sage.

—— (1989) "The real and unrealized contributions of quantitative sociology," *American Sociological Review* 54: 447–460.

Blau, P. (1995) "A circuitous path to macrostructural theory," *Annual Review of Sociology* 21: 1–19.

Blumer, H. (1931) "Science without concepts," *American Journal of Sociology* 36: 515–533.

—— (1954) "What is wrong with social theory?" *American Sociological Review* 19: 3–10.

—— (1969) *Symbolic Interactionism: Perspective and Method*, Englewood Cliffs NJ: Prentice Hall.

Bode, B. (1989) *No Bells to Toll: Destruction and Creation in the Andes*, New York: Scribners.

Bohle, H. (1993) "Vulnerability, hunger and famines," *GeoJournal* 30: 2–20.

Bolin, R. (1988) "Response to natural disaster," in M. Lystad (ed.) *Mental Health Responses to Mass Emergencies: Theory and Practice*, New York: Brunner/Mazel.

—— (1994) "Postdisaster sheltering and housing: social processes in response and recovery," in R. Dynes and K. Tierney (eds) *Disasters, Collective Behavior and Social Organization*, Newark DE : University of Delaware Press.

Bolton, P. (1986) "Natural hazards and industrial crises: emergency management considerations," unpublished paper.

Bosworth, S. and Kreps, G. (1986) "Structure as process: organization and role," *American Sociological Review* 51: 699–716.

Bottomore, T. and Nisbet, R. (eds.) (1979) *A History of Sociological Analysis*, New York: Basic Books.

Britton, N. (1987) "Towards a reconceptualization of disaster for the enhancement of social preparedness," in R. Dynes, B. De Marchi and C. Pelanda (eds) *Sociology of Disaster*, Milan, Italy: Franco Angeli.

Brook, D. (1992) "Policy in response to geohazards: lessons from the developed world?" in G. McCall, D. Laming, and S. Scott (eds) *Geohazards: Natural and Man-Made*, London: Chapman and Hall.

Brouillette, J. and Quarantelli, E. (1971) "Types of patterned variation in bureaucratic adaptation to organizational stress," *Sociological Inquiry* 41: 39–46.

Brown, M. and Goldin, A. (1973) *Collective Behavior: A Review and Reinterpretation of the Literature*, Pacific Palisades CA: Goodyear.

Brown, P. and Ferguson, F. (1995) "Making a big stink: women's work, women's relationships and toxic waste activists," *Gender and Society* 9: 145–172,

Buchanan-Smith, M. and Davies, S. (1995) *Famine Early Warning and Response: The Missing Link*, London: Intermediate Technology Publications.

Bunnin, N. and Tsu-James, E. (eds) (1996) *The Blackwell Companion to Philosophy*, Cambridge MA/Oxford: Blackwell.

Burchell, G., Gordon, C., and Miller, P. (eds) (1991) *The Foucault Effects: Studies in Governmentality*, London: Harvester Wheatsheaf.

Burns, T. (1958) "The forms of conduct," *American Journal of Sociology* 64: 137–151.

Burton, I., Kates, W., and White, G. (1978) *The Environment as Hazard*, New York: Oxford University Press.

Buttel, F. (1987) "Environmental sociology: a new paradigm?," *American Sociologist* 13: 252–256.

Bynagle, H. (1997) *Philosophy: A Guide to the Reference Literature*, Englewood CO: Libraries Unlimited.

Cable, S. and Cable, C. (1995) *Environmental Problems/Grassroots Solutions: The Politics of Environmental Conflict*, New York: St. Martin's.

Campbell, C. (1996) "Forest, field and factory: changing livelihood strategies in two extractive reserves in the Brazilian Amazon," unpublished PhD dissertation, University of Florida

Carr, L. (1932) "Disasters and the sequence-pattern concept of social change," *American Journal of Sociology* 38: 207–218.

Castel, R. (1991) "From dangerousness to risk," in G. Burchell, C. Gordon, and P. Miller (eds) *The Foucault Effect: Studies in Governmentality*, London: Harvester Wheatsheaf.

Castells, P. (1991) "International decade for natural disaster reduction," *UNDRO NEWS* (July/August): 19–20.

Catton, W. and Dunlap, R. (1978) "Environmental sociology: a new paradigm," *American Sociologist* 13: 41–49.

Clausen, L. (1978) *Tausch: Entwarfe zu einer Soziologischen Theorie*, München: Kasel-Verlag.

—— (1983) "Ubergang zum untergang: Skizze eines makrosoziologisches proze modells der katastrophe," *Zivilschutz-Forschung* 14: 41–79.

—— (1988) *Produktive Arbeit, Destruktive Arbeit: Soziologische Grundlagen*, Berlin: Walter de Gruyter.

—— (1992) "Social differentiation and the long-term origin of disasters," *Natural Hazards* 6: 181–190.

Clausen, L., Conlon, P., Jager, W., and Metreveli, S. (1978) "New aspects of the sociology of disaster: a theoretical note," *Mass Emergencies* 3: 61–65.

Clement, R. (1989) "The characteristics of risks of major disasters," *Proceedings of the Royal Society*, 242: 439–459.

Clough, P. (1988) "The movies and social observation: reading Blumer's movies and conduct," *Symbolic Interaction* 11: 85–97,

Cochrane, H. C. (1975) *Natural Hazards and Their Distributive Effect*, Boulder CO: Institute of Behavioral Science.

Cohen, M. (1996) "Risk society, ecological modernization, and declining public confidence in science," Working Paper ERC 96–7. Alberta, Canada: Environmental Risk Management, University of Alberta.

Cohen, M., March J., and Olsen J. (1972) "A garbage can model of organizational choice," *Administrative Science Quarterly* 17: 1–25.

Cole, S. (1992) *Making Science: Between Nature and Society*, Cambridge MA: Harvard University Press.

Collins, R. (1975) *Conflict Sociology*, New York: Academic Press

—— (1989) "Sociology: proscience or antiscience," *American Sociological Review* 54: 124–139.

Cook, L. (1981) *Demographic Collapse: Indian Peru, 1520–1620*, Cambridge: Cambridge University Press.

Copans, J. (ed.) (1975) *Secheresses et Famines du Sahel*, Paris: Maspero (in French).

Couch, S. and Kroll-Smith, R. (1994) "Environmental controversies, interactional resources, and rural communities: siting versus exposure disputes," *Rural Sociology* 59: 25–41.

Coulon, A. (1995) *Ethnomethodology* Thousand Oaks CA: Sage.

Craib, I. (1992) *Anthony Giddens*, London: Routledge.

Crisis Research Center (1996) *The Hercules Air Crash: Individuals, Organizations, Systems*, Leiden: Crisis Research Center.

Crozier, M. and Friedberg, E. (1979) *Macht und Organisation*, Berlin: Athenaum. (in German).

Cuny, F. (1983) *Disasters and Development*, Oxford: Oxford University Press.

Cutter, S. Tiefenbacher, J., and Solecki, W. (1992) "En-gendered fears: femininity and technological risk perception," *Industrial Crisis Quarterly* 6: 5–22.

Dahl, V. (1989) *Tolkovi Slovar Zhivoao Velikorusskogo Yazika*, Moskva: Russkiy Lazik (in Russian).

Dahlhamer, J. and Tierney, K. (1996) "Winners and losers: predicting business recovery following the Northridge earthquake," Preliminary Paper # 243. Newark DE: Disaster Research Center, University of Delaware.

Davis, I. (ed.) (1981) *Disasters and the Small Dwelling*, Oxford: Pergamon Press.

Deng, F. and Minear, L. (1992) *The Challenges of Famine Relief; Emergency Operations in the Sudan*, Washington DC: Brookings Institution.

Denzin, N. (1992) *Symbolic Interaction and Cultural Studies*, Oxford: Blackwell.

Devereux, S. (1993) *Theories of Famine*, New York: Harvester Wheatsheaf.

Dickens, D. and Fontana, A. (eds) (1994) *Postmodernism and Social Inquiry*, New York: Guilford.

Doherty, J., Graham, E., and Malek, M. (eds) (1992) *Postmodernism and the Social Sciences*, New York: St. Martin's Press.

Dombrowsky, W. (1981) "Another step toward a social theory of disaster," Preliminary Paper # 70. Newark DE: Disaster Research Center, University of Delaware.

—— (1983) "Solidarity during snow-disasters," *International Journal of Mass Emergencies and Disasters* 1: 189–205.

—— (1985) "Vom 'stage model' zumcopability profile: Katastrophensoziologische modellbildung in praktischer absicht," in L. Clausen and W. Dombrowsky (eds) *Einfuhrung in die Soziologie der Katastrophen*, Bonn, Germany: Bundesamt Für Zivilschultz (in German).

—— (1989) *Katastrophe und Katastrophenschutz: Eine Soziologische Analyse*, Wiesbaden: Deutscher Universitats-Velag.

—— (1993) "The social dimensions of warning and the transition from folk wisdom to laymanship," in J. Nemee (ed.) *Prediction and Perception of Natural Hazards*, Amsterdam: Kluwer Academic Publishers.

—— (1995a) "Again and again: Is a disaster what we call a disaster? Some conceptual notes on conceptualizing the object of disaster sociology," *International Journal of Mass Emergencies and Disasters* 13: 241–254.

—— (1995b) "Debate–Test–Dummy: a reaction to Hewitt's reaction paper," *International Journal of Mass Emergencies and Disasters* 13: 347–348.

Doughty, P. (1971) "From disaster to development," *Americas* 23: 23–35.

—— (1986) in A. Oliver-Smith and A. Hansen (eds) *Natural Disasters and Cultural Responses*, Williamsburg VA: College of William and Mary.

Douglas, J. (ed.) (1970) *Everyday Life: Toward the Reconstruction of Sociological Knowledge*, Chicago: Aldine.

Douglas, M. (1992) *Risk and Blame: Essays in Cultural Theory*, New York: Routledge.

Douglas, M. and Wildavsky, A. (1982) *Risk and Culture: An Essay on the Selection of Technical and Environmental Danger*, Berkeley CA: University of California Press.

Dove, M. and Khan, M. (1995) "Competing constructions of calamity: the April 1991 Bangladesh cyclone," *Population and Environment: A Journal of Interdisciplinary Studies* 16: 445–471.

Downs, R., Kerner, D., and Reyna, S. (eds) (1991) *The Political Economy of African Famine*, Philadelphia: Gordon and Breach Science Publishers.

Drabek, T. (1986) *Human System Response to Disaster: An Inventory of Sociological Findings*, New York: Springer-Verlag.

—— (1987) "Emergent structures," in R. Dynes, B. De Marchi, and C. Pelanda (eds) *Sociology of Disaster*, Milan, Italy: Franco Angeli.

—— (1989a) "Disasters as nonroutine social problems," *International Journal of Mass Emergencies and Disasters* 7: 253–264.

—— (1989b) "Taxonomy and disaster: theoretical and applied issues," in G. Kreps (ed.) *Social Structure and Disaster*, Newark DE: University of Delaware Press.

Drabek, T. and Haas, E. (1969) "Laboratory simulation of organizational stress," *American Sociological Review* 34: 223–238.

Drabek, T. and Quarantelli, E. (1967) "Scapegoats, villains and disasters," *Transaction* 4: 12–17.

Dreze, J. and Sen, A. (eds) (1990) *The Political Economy of Hunger*, Oxford: Clarendon Press.

Dubhashi, P. (1992) "Drought and development," *Economic and Political Weekly* (India) 27: A27–A36.

Dubin, R. (1978) *Theory Building*, New York: Free Press.

Duncan, O. (1964) "Social organization and the ecosystem," in R. Faris (ed.) *Handbook of Sociology*, Chicago: Rand McNally.

Dunlap, R. and Catton, W. (1983) "What environmental sociologists have in common (whether concerned with 'built' or 'natural' environments)," *Sociological Inquiry* 53: 113–135.

Durkheim, E. (1964) *The Division of Labor in Society*, New York: Free Press.

—— (1995) *The Elementary Forms of Religious Life*, New York: Free Press.

Dyer, C. (1995) "An analysis of the variability of institutional and cultural reactions to the impact of Hurricane Andrew on the fisheries of Florida versus Louisiana," unpublished paper.

Dynes, R. (1974) *Organized Behavior in Disaster*, Newark DE: Disaster Research Center, University of Delaware.

—— (1988) "Cross-cultural and international research: sociology of disaster," *International Journal of Mass Emergencies and Disasters* 6: 101–129.

—— (1993) "Disaster reduction: the importance of adequate assumptions about social organization," *Sociological Spectrum* 13: 175–192.

—— (1994a) "Conceptualizacion del desastre en formas produtivas para la investigacion en ciencias sociales," in A. Lavell (ed.) *Al Norte del Rio Grande Ciencias Sociales, Desastre: Una Perspectiva Norteamerican*, Columbia: LA RED (in Spanish).

—— (1994b) "Community emergency planning: false assumptions about social organization," *International Journal of Mass Emergencies and Disasters* 12: 141–158.

Dynes, R. and Drabek, T. (1994) "The structure of disaster research: its policy and disciplinary implications," *International Journal of Mass Emergencies and Disasters* 12: 5–23.

Dynes, R. and Quarantelli, E. (1968) "Redefinition of property norms in community emergencies," *International Journal of Mass Emergencies and Disasters* 3: 100–112.

—— (1976) "Community conflict: its absence and its presence in natural disasters," *Mass Emergencies* 1: 139–152.

—— (1980) "Helping behavior in large-scale disasters," in D. Smith and J. McAuley (eds) *Participation in Social and Political Activities*, San Francisco: Jossey-Bass.

Dynes, R., Quarantelli, E., and Wenger, D. (1990) *Individual and Organizational Response to the 1985 Earthquake in Mexico City, Mexico*, Newark DE: Disaster Research Center, University of Delaware.

Ebert, C. (1988) *Disasters: Violence of Nature and Threats by Man*, Dubuque: Kendall/Hunt.

Edelstein, M. (1991) "Ecological threats and spoiled identities: radon gas and environmental stigma," in S. Couch and J. Kroll-Smith (eds) *Communities at Risk*, New York: Peter Lang.

Eder, K. (1996) *The Social Construction of Nature: A Sociology of Ecological Enlightenment*, Thousand Oaks CA: Sage.

Ehrlich, A. and Birks, J. (1990) *Hidden Dangers: Environmental Consequences of Preparing for War*, San Francisco: Sierra Club Books.

Ekman, P. and Davidson, R. (eds) (1994) *The Nature of Emotion: Fundamental Questions*, New York: Oxford University Press.

El-Sabah, M. and Murty, T. (eds) (1988) *Natural and Man-Made Hazards: Proceedings of the International Symposium Held at Rimouski, Quebec, Canada, 3–9 August 1985*, Dordrecht: D. Reidel.

Elias, N. (1978) *What is Sociology?*, London: Hutchinson.

—— (1983) "Gedanken uber die grosse evolution. Zwei fragmente," in *Engagement und Distanzierung. Arbeiten zur Wissenssoziologie.* Frankfurt: Suhrkamp (in German).

Endleman, R. (1952) "An approach to the study of disaster," unpublished paper.

Erikson, K. (1976) *Everything in Its Path: Destruction of Community in the Buffalo Creek Flood*, New York: Simon and Schuster.

—— (1994) *A New Species of Trouble: Explorations in Disaster, Trauma and Community*, New York: Norton.

Fay, S. (1996) *The Collapse of Barings*, New York: Norton.

Featherstone, M. (1995) *Undoing Culture: Globalization, Postmodernism, and Identity*, Thousand Oaks CA: Sage.

Feder, B. (1993) "Winners as well as losers in the Great Flood of '93," *New York Times* August 15: F5.

Field, J. (ed.) (1993) *The Challenge of Famine: Recent Experience, Lessons Learned*, West Hartford CT: Kumarina Press.

Flaherty, M. (1993) "Conceptualizing variations in the experience of time," *Sociological Inquiry* 63: 394–405.

Forbes, L. and Sells, L. (1996) "Reorganizing the woman/nature connection," *Organization and Environment* 10: 20–22.

Forrest, T. (1978) "Group emergence in disasters," in E. Quarantelli (ed.) *Disaster: Theory and Research*, Beverly Hills CA: Sage.
Foster, H. (1990) *Disaster Mitigation for Planners: The Preservation of Life and Property*, New York: Springer Verlag.
Fothergill, A. (1996) "Gender, risk and disaster," *International Journal of Mass Emergencies and Disasters* 14: 33–56.
Foucault, M. (1975) *Surveiller et Punir*, Paris: Gallimard (in French).
—— (1991) "Governmentality," in G. Burchell, C. Gordon, and P. Miller (eds) *The Foucault Effect: Studies in Governmentality*, London: Harvester Wheatsheaf.
Fowler, B. (1997) *Pierre Bourdieu and Cultural Theory*, Thousand Oaks CA: Sage.
Franks, D. and McCarthy, E. (eds) (1989) *The Sociology of Emotions: Original Essays and Research Papers*, Greenwich: JAI Press.
Freudenburg, W. and Gramling, R. (1989) "The emergence of environmental sociology: contributions of R. E. Dunlap and W. R. Catton, Jr," *Sociological Inquiry* 59: 439–452.
Freudenburg, W. and Jones, T. (1991) "Attitudes and stress in the presence of a technological risk: a text of the Supreme Court hypothesis," *Social Forces* 69: 1143–1168.
Freudenburg, W. and Pastor, S. (1992) "Public response to technological risks: toward a sociological perspective," *Sociological Quarterly* 33: 389–412.
Fritz, C. (1961) "Disasters," in R. Merton and R. Nisbet (eds) *Social Problems,* New York: Harcourt Brace.
—— (1968) "Disasters," in *International Encyclopedia of the Social Sciences, Volume III*, New York: Macmillan.
Fritz, C. and Williams, H. (1957) "The human being in disasters: a research perspective," *Annals* 309: 42–51.
Furet, F. (ed.) (1989) *Unanswered Questions: Nazi Germany and the Genocide of the Jews*, New York: Schocken.
Gallie. W. (1955) "Essentially contested concepts," *Proceedings of the Aristotelian Society* 56: 167–198.
Geipel, R. (1982) *Disasters and Reconstruction: The Friuli Earthquakes of 1976*, London: Allen and Unwin.
Geertz, C. (1973) *The Interpretation of Cultures*, New York: Basic Books.
Giddens, A. (1984) *The Constitution of Society: Outline of the Theory of Structuration*, Berkeley: University of California Press.
—— (1987) *Social Theory and Modern Sociology*, Stanford: Stanford University Press.
—— (1990) *The Consequences of Modernity*, Cambridge: Polity Press.
—— (1991) *Modernity and Self-Identity*, Cambridge: Polity Press.
Gilbert, C. (1991) "Politique et complexité: les crises sans ennemi," in *Colloque International: Le Cadre Theorique de la Gestion des Crises dans les Societes Complexes: Etat de la Question*, Grenoble, France: CRISE (in French).
—— (1992) "The nature of politics in emergency situations," unpublished paper.
—— (1995) "Studying disaster: a review of the main conceptual tools," *International Journal of Mass Emergencies and Disasters* 13: 231–240.
Glantz, M. (1976) (ed.) *The Politics of Natural Disaster: The Case of the Sahel Drought*, New York: Praeger.
Glaser, B. and Strauss, A. (1967) *The Discovery of Grounded Theory*, Chicago: Aldine.

Gleick, J. (1987) *Chaos: Making a New Science*, New York: Viking.
Gosden, C. (1994) *Social Being and Time*, Oxford: Blackwell.
Gregersen, H. and Sailer, L. (1993) "Chaos theory and its implications for social science research," *Human Relations* 46: 777–802.
Gregor, A. (1971) *An Introduction to Metapolitics: A Brief Inquiry Into the Conceptual Language of Political Science*, New York: Free Press.
Griffin, D. (1995) from dgriffin@students.wisc.edu.
Gurvitch, G. (1964) *The Spectrum of Social Time*, Dordrecht: Reidel.
Gusfield, J. (1975) *Community*, New York: Harper.
Guastello, S. (1995) *Chaos, Catastrophe, and Human Affairs*, Mahwak NJ: Lawrence Erlbaum.
Haas, J. and Drabek, T. (1973) *Complex Organizations*, New York: Macmillan.
Hacking, I. (1990) *The Taming of Chance*, Cambridge: Cambridge University Press.
Hancock, G. (1989) *Lords of Poverty*, London: Mandarin.
Harre, R. (1986) *The Social Construction of Emotions*, New York: Blackwell.
Harrell-Bond, B. (1986) *Imposing Aid*, Oxford: Oxford University Press.
Harris, M. (1979) *Cultural Materialism: The Struggle for a Science of Culture*, New York: Random House.
Harrison, G. (1988) *Famine*, New York: Oxford.
't Hart, P. (1993) "Symbols, rituals and power: the lost dimensions of crisis management," *Journal of Contingencies and Crisis Management* 1: 36–50.
Hartmann, B. and Boyce, J. (1983) *A Quiet Violence: View From a Bangladesh Village*, London: Zed.
Harvey, D. (1996) *Justice, Nature and the Geography of Difference*, Cambridge MA: Blackwell.
Haynor, A. (1990) "In defense of universal theory," *Perspectives* 13: 16–36.
"Hazards Assessment Update" (1996), *Natural Hazards Observer* 21: 6–7.
Hempel, C. (1952) *Fundamentals of Concept Formation in Empirical Research*, Chicago: University of Chicago Press.
Hermann, C. (1972) *International Crises: Insights from Behavioral Research*, New York: Free Press.
Hewitt, K. (1983a) "The idea of calamity in a technocratic age," in K. Hewitt (ed.) *Interpretations of Calamity: From the Viewpoint of Human Ecology*, London: Allen and Unwin.
—— (1983b) *Interpretations of Calamity From the Viewpoint of Human Ecology* , Boston: Allen and Unwin.
—— (1987) "Risks and emergencies in Canada: a national overview," *Ontario Geography* 29: 1–36.
—— (1992) "Mountain hazards," *GeoJournal* 27: 47–60.
—— (1994a) "When the great planes came and made ashes of our city . . . Towards an oral geography of the disasters of war," *Antipode* 26: 1–34.
—— (1994b) "Hidden damages, shadow risks: making the social space of disasters visible," in *Proceedings: Seminario Internactional Sociedad y Prevenction de Desastres*, Mexico City; Autonomous University.
—— (1995) "Excluded perspectives in the social construction of disaster," *International Journal of Mass Emergencies and Disasters* 13: 317–339.

Hewitt, K. and Burton, I. (1971) *The Hazardousness of a Place: A Regional Ecology of Damaging Events*, Toronto: University of Toronto Press.

Hindess, B. (1977) *Philosophy and Methodology in the Social Sciences*, Atlantic Highlands NJ: Humanities Press.

Hohenemser, C., Kasperson, C., and Kates, R. (1985) in R. Kates, C. Hohenemser, and J. Kasperson (eds) *Perilous Progress; Managing the Hazards of Technology*, Boulder CO: Westview.

Holling, C. (1994) "An ecologist view of the Malthusian conflict," in K. Lindahl-Kiessling and H. Landberg (eds) *Population, Economic Development and the Environment*, New York: Oxford University Press.

Homans, G. (1967) *The Nature of Social Science*, New York: Harcourt Brace and World.

Horlick-Jones. T. (1995) "Modern disasters as outrage and betrayal," *International Journal of Mass Emergencies and Disasters* 13: 305–316.

—— (1996) "The problem of blame," in C. Hood and D. Jones (ed.) *Accident and Design: Contemporary Debates in Risk Management*, London: UCL Press.

Horlick-Jones, T., Fortune, J., and Peters, G. (1991a) "Measuring disaster trends. Part I: Some observations on the Bradford fatality scale," *Disaster Management* 3: 144–148.

—— (1991b) "Measuring disaster trends. Part II: Statistics in the underlying processes," *Disaster Management* 4: 41–45.

Horowitz, M. and Salem-Murdock, M. (1987) "The political economy of desertification in White Nile Province, Sudan," in P. Little, M. Horowitz, and R. Nyerges (eds) *Lands at Risk in the Third World: Local Level Perspectives*, Boulder CO: Westview Press.

Huff, T. (1973) "Theoretical innovation in science: the case of William F. Ogburn," *American Journal of Sociology* 79: 261–277.

Ingold, T. (1992) "Culture and the perception of the environment," in E. Croll and D. Parkin (eds) *Bush Base: Forest Farm*, London: Routledge.

International Atomic Energy Agency (1986) *Summary Report on the Post-accident Review Meeting on the Chernobyl Accident*, Vienna: IAEA.

Jager, W. (1977) *Katastrophe und Gesellschaft: Grundlegungen und Kritik von Modellen der Katastrophensoziologie*, Darmstadt: Luchterhand.

Johnson, N. (1988) "Fire in a crowded theater: a descriptive analysis of the emergence of panic," *International Journal of Mass Emergencies and Disasters* 6: 7–26.

Johnson, M. and Hufbauer, K. (1982) "Sudden infant death syndrome as a medical research problem since 1945," *Social Problems* 30: 65–81,

Johnson, N., Feinberg, W., and Johnston, D. (1994) "Microstructure and panic: the impact of social bonds on individual action in collective flight from the Beverly Hills supper club fire," in R. Dynes and K. Tierney (eds) *Disasters, Collective Behavior, and Social Organization*, Newark DE: University of Delaware Press.

Johnston, D. and Johnson, N. (1989) "Role expansion in disaster: an investigation of employee behavior in a nightclub fire," *Sociological Focus* 22: 39–51.

Johnston, H. and Klandermans, B. (eds) (1995) *Social Movements and Culture*, Minneapolis MN: University of Minnesota Press.

Jones, D. (ed.) (1993) "Environmental hazards: the challenge of change," *Geography* 63: 161–198.

Karplus, W. (1992) *The Heavens Are Falling: The Scientific Prediction of Catastrophes in Our Time*, New York: Plenum Press.

Kartashov, A. (1991) "Stand up and fight! Dealing with disasters: Soviet experience," unpublished paper.

Katastrophen-Vorschrift (1988), Bonn: Deutsches Rotesse Kreuz (in German).

Keen, D. (1994) *The Benefits of Famine: A Political Economy of Famine and Relief in Southwestern Sudan, 1983–1989* Princeton NJ: Princeton University Press.

Kent, G. (1984) *The Political Economy of Hunger: The Silent Holocaust*, New York: Praeger.

Keown-McMullan, C. (1997) "Crisis: when does a molehill become a mountain?" *Disaster Prevention and Management* 6: 4–10.

Kiel, D. and Elliott, E. (1996) *Chaos Theory in the Social Sciences: Foundations and Application*, Ann Arbor MI: University of Michigan.

Killian, L. (1954) "Some accomplishments and some needs in disaster study," *Journal of Social Issues* 10: 66–72.

Kitayama, S. and Markus, H. (eds) (1994) *Emotion and Culture: Empirical Studies of Mutual Influence*, Washington DC: American Psychological Association.

Koehler, G. (ed.) (1995) *What Disaster Response Management Can Learn From Chaos Theory: Conference Proceedings*, Sacramento CA: California Research Bureau (also available at http://library.ca./gov/CRB/over_toc.html).

Kogon, E. (1958) *The Theory and Practice of Hell*, New York: Berkeley Publishing.

Kohn, G. (ed.) (1995) *Encyclopedia of Plague and Pestilence*, New York: Facts on File.

Kolata, A. (1993) *The Tiwanaku: Portrait of an Andean Civilization*, Cambridge: Blackwell.

Kovoor-Misra, S. (1995) "A multidimensional approach to crisis preparation for technical organizations: some critical factors," *Technological Forecasting and Social Change* 48: 143–160.

Kreps, G. (1973) *Decision Making Under Conditions of Uncertainty: Civil Disturbance and Organizational Change in Urban Police and Fire Departments*, Newark DE: Disaster Research Center, University of Delaware.

—— (1978) "The organization of disaster response: Some fundamental theoretical issues," in E. Quarantelli (ed.) *Disasters: Theory and Research*, Beverly Hills CA: Sage.

—— (1983) "The organization of disaster response: core concepts and processes," *International Journal of Mass Emergencies and Disasters* 1: 439–465.

—— (1984) "Sociological inquiry and disaster research," *Annual Review of Sociology* 10: 309–330.

—— (1985) "Disaster and the social order," *Sociological Theory* 3: 49–65.

—— (1989a) "Disasters and the social order," in G. Kreps (ed.) *Social Structure and Disaster*, Newark DE: University of Delaware Press.

—— (1989b) *Social Structure and Disaster*, Newark DE: University of Delaware Press.

—— (1989c) "Future directions in disaster research: the role of taxonomy," *International Journal of Mass Emergencies and Disasters* 7: 215–241.

—— (ed.) (1989d) "The boundaries of disaster research: taxonomy and comparative research," *International Journal of Mass Emergencies and Disasters* 7 (Special Issue): 213–431

—— (1991) "Answering organizational questions: a brief for structural codes," in G. Miller (ed.) *Studies in Organizational Sociology* , Greenwich CT: JAI Press.

—— (1993) "Disaster as systemic event and social catalyst: a clarification of subject matter," unpublished paper.

—— (1995a) "Disaster as systemic event and social catalyst: a clarification of subject matter," *International Journal of Mass Emergencies and Disasters* 13: 255–284.

—— (1995b) "Excluded perspectives in the social construction of disaster: a response to Hewitt's critique," *International Journal of Mass Emergencies and Disasters* 13: 349–351.

Kreps, G. and Bosworth, S. (1993) "Disaster, organizing and role enactment: a structural approach," *American Journal of Sociology*, 99: 428–463.

Kreps, G. and Bosworth S. with Mooney, J., Russell, S., and Meyers, K. (1994) *Organizing Role Enactment and Disaster: A Structural Theory*, Newark DE: University of Delaware Press.

Kreps, G. and Drabek, T. (1996) "Disasters are nonroutine social problems," *International Journal of Mass Emergencies and Disasters* 14: 129–153.

Kriterii Otsenki Ecologicheskoi Obstanovki Territoriy dlia Viyavilenai Zon Chrezvichanoi Ecologicheskoi Situatsii i Zon Ecologicheskogo Bedstviya (1994), Moskva: Zeleniy Mir (in Russian).

Kroeber, A. and Kluckhohn, C. (1952) *Culture: A Critical Review of Concepts and Definitions*, New York: Vintage Books.

Kroll-Smith, J. and Couch, S. (1990a) *The Real Disaster is Above Ground: A Mine Fire and Social Conflict*, Lexington: University of Kentucky Press.

—— (1990b) "Sociological knowledge and the public at risk: a 'self-study' of sociology, technological hazards and moral dilemmas," *Sociological Practice Review* 1: 120–127.

—— (1991) "What is a disaster? An ecological-symbolic approach to resolving the definitional debate," *International Journal of Mass Emergencies and Disasters* 9: 355–366.

Kroll-Smith, S. and Floyd, H. (forthcoming) *Bodies in Protest: Environmental Illness and the Struggle over Medical Knowledge*, New York: New York University Press.

Kuhn, T. (1970) *The Structure of Scientific Revolutions*, Chicago: University of Chicago Press.

Lagadec, P. (1988) *Etats d'Urgence*, Paris: Le Seuil (in French).

—— (1991) *La Gestion des Crises*, Paris: McGraw-Hill (in French).

LaPorte, T. and Consolini, P. (1991) "Working in practice but not in theory: theoretical challenges of 'high reliability organizations,'" *Journal of Public Administration Research and Theory* 1: 19–47.

Latour, B. (1995) *Sociologie des Sciences, Analyse des Risques Collectifs et des Situations de Crise*, Grenoble, France: CNRS (in French).

Lauristin, M. (1996) "Estonia ferry disaster and the Estonian Information Service," in I. Johansson and E. Skoglund (eds) *Crisis Management at the National Level*, Stockholm: Modin Tryck AB.

Lebow, R. (1981) *Between War and Peace: The Nature of International Crisis*, Baltimore: Johns Hopkins University Press.

Lenski, G. (1988) "Rethinking macrosociological theory," *American Sociological Review* 53: 163–171.
Levine, A. (1982) *Love Canal: Science, Politics, and People*, Lexington MA: Lexington Books.
Lewis, J. (1987) *Vulnerability and Development–and the Development of Vulnerability: A Case for Management*, London: Plenum International.
Lewis, M. and Haviland, J. (eds) (1993) *Handbook of Emotions*, New York: Guilford Press.
Lifton, R. (1974) *Death in Life: Survivors of Hiroshima*, New York: Random House.
Lifton, R. and Markusen, E. (1990) *The Genocidal Mentality*, New York: Basic Books.
Lindblom, C. (1977) *Politics and Markets: The World's Political-Economic Systems*, New York: Basic Books.
Lindell, M. (1997) "Special issue: natural hazard mitigation in the United States," *International Journal of Mass Emergencies and Disasters* 15: 432–559.
Lindell, M. and Perry, R. (1996) "Identifying and managing conjoint threats: earthquake-induced hazardous materials releases," *Journal of Hazardous Materials* 50: 31–46.
Liverman, D. (1990) "Vulnerability to global environmental change," in R. Kasperson (ed.) *Understanding Global Environmental Change: The Contributions of Risk Analysis and Management*, Worcester MA: Clark University.
—— (1993) "Drought impacts in Mexico: climate, agriculture, technology and land tenure in Sonora and Puebla," *Annals of the Association of American Geographers* 80: 49–72.
Lofland, J. (1989) "Consensus movements: city twinning and derailed dissent in the American eighties," *Research in Social Movements: Conflict and Change* 11: 163–196.
Luhmann, N. (1992) *The Differentiation of Society*, New York: Columbia University Press.
—— (1993) *Risk: A Sociological Theory*, New York: Walter de Gruyter.
—— (1995) *Social Systems*, Stanford CA: Stanford University Press.
Lyman, S. (1994) "The bequests of 20th Century sociology to the 21st Century," *Sociological Spectrum* 15: 209–225.
McAdam, D., Tarrow, S., and Tilly, C. (1996) "To map contentious politics," *Mobilization: An International Journal* 1: 17–34.
McCall, G., Laming, D., and Scott, S. (eds) (1992) *Geohazards: Natural and Man-made*, London: Chapman and Hall.
McCann, J. (1987) *From Poverty to Famine in Northeast Ethiopia: A Rural History 1900–1935*, Philadelphia: University of Pennsylvania Press.
McCarthy, J. and Wolfson, M. (1992) "Consensus movements, conflict movements and the cooptation of civic and state infrastructure," in A. Morris and C. Mueller (eds) *Frontiers in Social Movement Theory*, New Haven CT: Yale University Press.
McGrath, J. (1988) *The Social Psychology of Time: New Perspectives*, Newbury Park CA: Sage.
McGrath, J. and Kelly, J. (1986) *Time and Human Interaction: Toward a Social Psychology of Time*, New York: Guilford Press.
McKinney, J. (1966) *Constructive Typology and Social Theory*, New York: Appleton Century Crofts.

—— (1969) "Typification, typologies and sociological theory," *Social Forces* 48:1–12

Macksoud, M. (1992) "Assessing war trauma in children: a case study of Lebanese children," *Journal of Refugee Studies* 5: 1–15.

McQuail, D. (1993) *Media Performance: Mass Communication and the Public Interest*, London: Sage.

March, J. and Olsen, J. (1979) *Ambiguity and Choice*, Bergen, Norway: Universitatsforlaget.

Margolis, H. (1996) *Dealing With Risk: Why the Public and the Experts Disagree on Environmental Issues*, Chicago: University of Chicago Press.

Margolis, M. and Murphy, M. (1995) *Science, Materialism and the Study of Culture*, Gainesville FL: University Press of Florida.

Marks, E. and Fritz, C. (1954) *Human Reactions in Disaster Situations*, Chicago: National Opinion Research Center, University of Chicago.

Marples, D. (1988) *The Social Impact of the Chernobyl Disaster*, New York: St. Martin's Press.

Marx, K. (1961) *Kapital. Kritika Politicheski Ekonomii*, Moskva: Politizdat (in Russian).

Marx, K. (1964) *Selected Writings in Sociology and Social Philosophy*, London: McGraw Hill.

Marx, G. and McAdam, D. (1994) *Collective Behavior and Social Movements: Process and Structure*, Englewood Cliffs NJ: Prentice Hall.

Maskrey, A. (1989) *Disaster Mitigation: A Community Based Approach*, Oxford: Oxfam.

May, P. and Williams, W. (1986) *Disaster Policy Implementation: Managing Programs Under Shared Governance*, New York: Plenum Press.

Medvedev, Z. (1990) *The Legacy of Chernobyl*, New York: Norton.

Melucci, A. (1996) *Challenging Codes: Collective Action in the Information Age*, Cambridge, England: Cambridge University Press.

Merchant, C. (1996) *Earthcare: Women and Environment*, New York: Routledge.

Merton, R. (1945) "Sociological theory," *American Journal of Sociology* 50: 462–473.

—— (1968) *Social Theory and Social Structure*, New York: Free Press.

—— (1972) "Insiders and outsiders: A chapter in the sociology of knowledge," *American Journal of Sociology* 78: 9–47.

Merton, R. and Kitt, A. (1950) "Contributions to the theory of reference group behavior," in R. Merton (ed.) *Continuities in Social Research*, Glencoe IL: Free Press.

Mestrovic, S. (1996) *Postemotional Society*, Thousand Oaks CA: Sage.

Meyer, E. and Poniatowska, E. (1988) "Documenting the earthquake of 1985 in Mexico City," *Oral History Review* 16: 1–31.

Mileti, D. and Fitzpatrick, C. (1993) *The Great Earthquake Experiment: Risk Communication and Public Action*, Boulder CO: Westview Press.

Mileti, D., Drabek, T., and Haas, J. (1975) *Human Systems and Extreme Environments*, Boulder CO: Institute of Behavioral Science, University of Colorado.

Milton, K. (1996) *Environmentalism and Cultural Theory: Exploring the Role of Anthropology in Environmental Discourse*, London: Routledge.

Mitchell, J. (1990) "Human dimensions of environmental hazards: complexity, disparity and the search for guidance," in A. Kirby (ed.) *Nothing to Fear: Risks and Hazards in American Society*, Tucson AZ: University of Arizona Press.

—— (1993) "Recent developments in hazards research: a geographer's perspective," in E. Quarantelli and K. Popov (eds) *Proceedings of the United States–Former Soviet Union Seminar on Social Science Research on Mitigation for and Recovery from Disasters and Large Scale Hazards*, Newark DE: Disaster Research Center, University of Delaware.

Mitroff, I. and Kilmann, R. (1984) *Corporate Tragedies* New York: Praeger.

Mol, A. and Spaargaren, G. (1993) "Environment, modernity, and the risk-society: the apocalyptic horizon of environmental reform," *International Sociology*, 8: 431–459.

Moore, H. (1956) "Toward a theory of disaster," *American Sociological Review* 21: 734–737.

—— (1964) *. . . And the Winds Blew*, Austin TX: Hogg Foundation, University of Texas.

Moore, J. (1996) *The UN and Complex Emergencies: Rehabilitation in Third World Transitions*, Geneva: UNRISD.

Morren, G. (1983) "A general approach to the identification of hazards," in K. Hewitt (ed.) *Interpretations of Calamity: From the Viewpoint of Human Ecology*, London: Allen and Unwin.

Morrison, D. and Hornbeck, K. (eds) (1976) *Collective Behavior: A Bibliography*, New York: Garland.

Morrow, B. and Enarson, E. (1996) "Hurricane Andrew through women's eyes: issues and recommendations," *International Journal of Mass Emergencies and Disasters* 14: 5–12.

Moseley, M., Feldman, R., and Ortloff, C. (1981) "Living with crises: human perceptions of process and time," in M. Nitecki (ed.) *Biotic Crises in Ecological and Evolutionary Time*, Princeton: Princeton University Press.

Mulkay, M. (1991) *Sociology of Science–A Sociological Pilgrimage*, Philadelphia: Open University.

National Governors' Association (1979) *Comprehensive Emergency Management*, Washington DC: Defense Civil Preparedness Agency.

National Research Council (1987) *Confronting Natural Disasters: An International Decade for Natural Hazards Reduction*, Washington DC: National Academy of Sciences.

Neal, D. (1997) "Reconsidering the phases of disasters," *International Journal of Mass Emergencies and Disasters* 15.

Nelson, B. (1984) *Making An Issue of Child Abuse: Political Agenda Setting for Social Problems*, Chicago: University of Chicago Press.

Nicholls, C. (1990) *Power: A Political History of the Twentieth Century*, New York: Oxford University Press.

Nikiforuk, A. (1993) *The Fourth Horsemen: A Short History of Epidemics, Plagues, Famine and Other Scourges*, New York: Evans.

Notfailschutz (1986) *Notfailschutz für die Umgebung des Kernraftwerkes Muhlhelm-Karlich. Ein Ratgeber für die Bevolkerung Ministerium des Innern und für Sport desLandes Rheinland Pflax*, Mainz: Innenministeerium (in German).

Nowotny, H. (1994) *Time: The Modern and Postmodern Experience*, Cambridge: Polity Press.

Nunez de la Pena, F. and Orozco, J. (1988) *El Terremoto: Una Version Correaida*, Guadalajara, Mexico: Iteso (in Spanish).
Offe, C. (1972) *Strukturprobleme des Kapitalistisches Staates: Aufsatze zur Politischen Soziologie*, Frankfurt: Suhrkamp Verlag.
O'Keefe, P. (1975) *Disastrous Relief*, London: War on Want.
O'Keefe, P., Westgate, K., and Wisner, B. (1976) "Taking the naturalness out of natural disaster," *Nature* 91: 260.
Oliver, P. (1989) "Bringing the crowd back in," in L. Kriesberg (ed.) *Research in Social Movements, Conflict and Change Volume 11*, Greenwich CT: JAI Press.
Oliver-Smith, A. (1986) "Responses to floods in the English countryside," in A. Oliver-Smith and A. Hansen (eds) *Natural Disasters and Cultural Responses*, Williamsburg VA: College of William and Mary.
—— (1992) *The Martyred City: Death and Rebirth in the Andes*, 2nd edition, Prospect Park: Waveland Press.
—— (1993) "Anthropological perspective in disaster research," in E. Quarantelli and K. Popov (eds) *Proceedings of the United States–Former Soviet Union Seminar on Social Science Research on Mitigation for and Recovery from Disasters and Large Scale Hazards*, Newark DE: University of Delaware.
—— (1994) "Peru's five hundred year earthquake: vulnerability in historical context," in A. Varley (ed.) *Disasters, Development and Environment*, London: Wiley.
—— (1996) "Anthropological research on hazards and disasters," *Annual Review of Anthropology*, 25: 303–328.
Owen, D. (ed.) (1997) *Sociology After Postmodernism*, Thousand Oaks CA: Sage.
Ozhegov, S. (1987) *Slovar Russkogo Yazika*, Moskva: Russkii (in Russian).
Painter, M. and Durham, W. (1995) *The Social Causes of Environmental Destruction in Latin America*, Ann Arbor MI: University of Michigan Press.
Palm, R. (1990) *Natural Hazards: An Interactive Framework for Research and Planning*, Baltimore: Johns Hopkins University Press.
Papineau, D. (1978) *For Science in the Social Sciences*, New York: St. Martin's Press.
Parsons, T. (1951) *The Social System*, London: Routledge and Kegan.
Peacock, W., Morrow, B., and Gladwin, H. (in press) *Ethnicity, Gender and the Political Ecology of Disasters: Hurricane Andrew and the Reshaping of a City*, Gainesville: University Press of Florida.
Peet, R. and Watts, M. (1993) " Development theory and environment in an age of market triumphalism," *Economic Geography* 69: 227–253.
Pelanda, C. (1981) "Disasters and sociosystemic vulnerability," Preliminary Paper # 68, Newark DE: Disaster Research Center, University of Delaware.
—— (1982a) *Disaster and Social Order: Theoretical Problems in Disaster Research*, Gorizia, Italy: Institute of International Sociology.
—— (1982b) "Disaster and sociosystemic vulnerability," in B. Jones and M. Tomazevic (eds) *Social and Economic Aspects of Earthquakes*, New York: Cornell University.
Perrow, C. (1984) *Normal Accidents: Living With High-Risk Technologies*, New York: Basic Books.
—— (1997) "Organizing for environmental destruction," *Organization and Environment* 10: 66–72.
Perry, R. (1982) *The Social Psychology of Civil Defense*, Lexington MA: D. C. Heath.

—— (1985) *Comprehensive Emergency Management: Evacuating Threatened Populations*, Greenwich CT: JAI Press.

—— (1989a) "Taxonomy and model building for emergency warning response," *International Journal of Mass Emergencies and Disasters* 7: 305–327.

—— (1989b) "Taxonomy, classification, and theories of disaster phenomena," in G. Kreps (ed.) *Social Structure and Disaster*, Newark DE: University of Delaware Press.

—— (1990) "Managing disaster response operations," in T. Drabek and G. Hoetmer (eds) *The Principles and Practice of Emergency Management*, Washington DC: International City Management Association.

Petak, W. (ed.) (1985) "Special issue on emergency management: a challenge for public administration," *Public Administration Review* 45: 1–172.

Phillips, B. (1990) "Gender as a variable in emergency response," in R. Bolin (ed.) *The Loma Prieta Earthquake: Studies of Short-Term Impact*, Boulder CO: Institute of Behavioral Science, University of Colorado.

Picou, J. and Gill, D. (1996) "The *Exxon Valdez* oil spill and chronic psychological stress," in F. Rice, R. Spies, D. Wolfe, and B. Wright (eds) *Proceedings of the Evos Symposium*, Alaska: American Fisheries Symposium.

Picou, J., Gill, D., and Cohen, M. (eds) (1997) *The* Exxon Valdez *Disaster: Readings on a Modern Social Problem*, Dubuque IA: Kendall/Hunt Publishing Company.

Plumwood, V. (1993) *Feminism and the Mastery of Nature*, London: Routledge.

Porfiriev, B. (1989) *Organizatsiya Upravleniya v Chrezvichainikh Situatsiyakh: Problemi Predotvrazcheniya i Sokrazcheniya Masshtabov Posledstviy Priorodnikh i Technologicheskigk Katastrof*, Moskva: Znanie (in Russian).

—— (1991) *Gosudarstvennoye Upravleniye v Chrezvichainikh Situatsiyakh: Analiz Metodologii i Problemi Organizatsii*, Moskva: Nauka Publ. (in Russian).

—— (1995) "Disaster and disaster areas: methodological issues of definition and delineation," *International Journal of Mass Emergencies and Disasters* 13: 285–304.

Porfiriev, B. and Quarantelli, E. (eds) (1996) *Social Science Research on Mitigation of and Recovery From Disasters and Large Scale Hazards in Russia*, Newark DE: Disaster Research Center, University of Delaware.

Porter, B. and Dunn, M. (1984) *The Miami Riot of 1980*, Lexington MA: Heath.

Powell, J., Rayner, R., and Finesinger, J. (1952) "Responses to disaster in American cultural groups," in *Symposium on Stress*, Washington DC: Army Medical Service Graduate School.

Prendergast, C. and Knottnerous, J. (1994) "Recent developments in the theory of social structure: introduction and overview," *Current Perspective in Social Theory, Supplement* 1:1–26.

Presidential Commission on the Space Shuttle *Challenger* Accident (1986) *Report to the President by the Presidential Commission on the Space Shuttle Challenger Accident* (5 volumes), Washington DC: Government Printing Office.

Prince, S. (1920) *Catastrophe and Social Change*, New York: Columbia University Press.

Pronovost-Giles, A. (1989) "The sociological study of time: historical landmarks," *Current Sociology* 37: 4–19.

Quarantelli, E. (1954) "The nature and conditions of panic," *American Journal of Sociology* 60: 267–275.

—— (1957) "The behavior of panic participants," *Sociology and Social Research* 41: 187–194.

—— (1966) "Organization under stress," in R. Brictson (ed.) *Symposium on Emergency Operations*, Santa Monica CA: Systems Development Corporation.

—— (1970a) "The community general hospital: Its immediate problems in disasters," *American Behavioral Scientist* 13: 380–391.

—— (1970b) "Emergent accommodation groups: beyond current collective behavior typologies," in T. Shibutani (ed.) *Human Nature and Collective Behavior*, Englewoods Cliffs NJ: Prentice Hall.

—— (1977) "Social aspects of disaster and their relevance to pre-disaster planning," *Disasters* 1: 98–107.

—— (ed.) (1978) *Disasters: Theory and Research*, London: Sage.

—— (1980) "The study of disaster movies: research problems, findings and implications," Preliminary Paper # 64, Newark DE: Disaster Research Center, University of Delaware.

—— (1981) "An agent specific or an all disaster spectrum approach to social-behavioral aspects of earthquakes," Preliminary Paper # 69, Newark DE: Disaster Research Center, University of Delaware.

—— (1982) "What is a disaster? An agent specific or an all disaster spectrum approach to socio-behavioral aspects of earthquakes," in B. Jones and M. Tomazevic (eds) *Social and Economic Aspects of Earthquakes*, Ithaca NY: Program in Regional and Urban Studies, Cornell University.

—— (1983) *Delivery of Emergency Medical Services in Disasters: Assumptions and Realities*, Newark DE : Disaster Research Center, University of Delaware.

—— (1985a) "An assessment of conflicting views on mental health: the consequences of traumatic events," in C. Figley (ed.) *Trauma and Its Wake: The Treatment of Post-Traumatic Stress*, New York: Brunner/Mazel.

—— (1985b) "What is a disaster? The need for clarification in definition and conceptualization in research," in B. Sowder (ed.) *Disasters and Mental Health: Selected Contemporary Perspectives*, Washington DC: US Government Printing Office.

—— (1986) "What should we study?' Presidential address to the International Sociological Association Research Committee on Disasters at the World Congress of Sociology in New Delhi, India.

—— (1986–1987) "Le jour ou le desastre frappera vous serez admirable," *Temps Strategique* 19: 75–80.

—— (1987a) "Disaster studies: an analysis of the social historical factors affecting the development of research in the area," *International Journal of Mass Emergencies and Disasters* 5: 285–310.

—— (1987b) "What should we study? Questions and suggestions for researchers about the concept of disasters," *International Journal of Mass Emergencies and Disasters* 5: 7–32.

—— (1989a) "Conceptualizing disasters from a sociological perspective," *International Journal of Mass Emergencies and Disasters* 7: 243–251.

—— (1989b) "Panel: What should we be studying?" in E. Quarantelli and C. Pelanda (ed.) *Proceedings Italy–United States Seminar: Preparations for, Responses to, and Recovery from Major Community Disasters*, Newark DE: Disaster Research Center, University of Delaware.

—— (1991) "Disaster response: generic or agent-specific?" in A. Kreimer and M. Munasinghe (eds) *Managing Natural Disasters and the Environment,* Washington DC: Environment Department, World Bank.

—— (1992a) "The case for a generic rather than agent specific agent approach to disasters," *Disaster Management* 2: 191–196.

—— (1992b) "Disaster research," in E. Borgatta and M. Borgatta (eds) *Encyclopedia of Sociology*, New York: Macmillan.

—— (1993a) "Community crises: an exploratory comparison of the characteristics and consequences of disasters and riots," *Journal of Contingencies and Crisis Management* 1: 67–78.

—— (1993b) "Disasters and catastrophes: their conditions in and consequences for social development," Preliminary Paper # 197, Newark DE: Disaster Research Center, University of Delaware.

—— (1993c) "Technological and natural disasters and ecological problems: similarities and differences in planning for and managing them," in *Memoria del Coloquio Internacional: El Reto De Desastres Technologicos y Ecologicos*, Mexico City, Mexico: Academia Mexicana de Ingenieria.

—— (1994) "Disaster studies: the consequences of the historical use of a sociological approach in the development of research," *International Journal of Mass Emergencies and Disasters* 12: 5–23.

—— (1995a) "Disasters are different, therefore planning for and managing them requires innovative as well as traditional behavior," Preliminary Paper # 221, Newark DE: Disaster Research Center, University of Delaware.

—— (1995b) "Drafts of a sociological disaster research agenda for the future: theoretical, methodological and empirical issues," unpublished paper.

—— (1995c) "Epilogue," *International Journal of Mass Emergencies and Disasters* 13: 361–364.

—— (1995d) "Patterns of sheltering and housing in US disasters," *Disaster Prevention and Management* 4: 43–58.

—— (1995e) "What is a disaster?" *International Journal of Mass Emergencies and Disasters* 13: 221–229.

—— (1996) "The future is not the past repeated: projecting disasters in the 21st Century from current trends," *Journal of Contingencies and Crisis Management* 4: 228–240.

—— (1997) "The Disaster Research Center field studies of organized behavior in the crisis time period of disasters," *International Journal of Mass Emergencies and Disasters* 15: 47–70.

Quarantelli, E. and Drabek, T. (1967) "Scapegoats, villains, and disasters," *Trans-Action* 4: 12–17.

Quarantelli, E. and Dynes, R. (1970) "Property norms and looting: their pattern in community crises," *Phylon* 31: 168–182.

—— (1977) "Response to social crisis and disaster," *Annual Review of Sociology* 3: 23–49.

Quarantelli, E. and Mozgovaya, A. (eds) (1994) *An Annotated Inventory of the Social Science Research Literature on Disasters in the Former Soviet Union and Contemporary Russia*, Newark DE: Disaster Research Center, University of Delaware.

Renee di Pardo, D., Novelo, V., Rodriguez, M., Calvo, B., Galvan, L., and Macias J. (1987) *Terremoto y Sociedad*, Tlalpan, Mexico: Centro de Investigaciones y Estudios Superiores en Antropologia Soxcial.

Richardson, B. (1994) "Crisis management and management strategy: time to 'loop the loop,' " *Disaster Prevention and Management* 3: 59–80.

Rifkin, J. (1980) *Entropy: A New World View*, New York: Viking Press.

Robinson, R. and Combs, A. (eds) (1995) *Chaos Theory in Psychology and Life Sciences*, Mahwah NJ: Lawrence Erlbaum.

Robinson, R., Franco H., Casterejon, R., and Bernard, H. (1986) "It shook again: the Mexico City earthquake of 1985," in V. Sutlive, N. Altshuler, M. Zamora, and V. Kerns (eds) *Natural Disasters and Cultural Responses*, Williamsburg VA: College of William and Mary.

Rocheleau, D. Thomas-Slayter, B., and Wangari, E. (eds) (1996) *Feminist Political Ecology: Global Issues and Local Experiences*, London: Routledge.

Rochford, E. and Blocker, T. (1991) "Coping with 'natural' hazards as stressors: the predictors of activism in a flood disaster," *Environment and Behavior* 23: 171–194.

Rosaldo, R. (1989) *Culture and Truth: The Remaking of Social Analysis*, Boston: Beacon Press.

Rosenthal, U. (1986) "Governmental decision-making in crisis situations: decisions at hectic moments," unpublished paper.

Rosenthal, U. and Kouzmin, A. (1993) "Globalizing an agenda for contingencies and crisis management," *Journal of Contingencies and Crisis Management* 1: 1–12.

Rosenthal, U., 't Hart, P., and Charles, M. (1989) "The world of crises and crisis management," in U. Rosenthal, M. Charles, and P. 't Hart (eds) *Coping With Crises: The Management of Disasters, Riots and Terrorism*, Springfield IL: Charles C. Thomas.

Rosenthal, U., Kouzmin, A., and 't Hart, P. (eds) (1989) *Coping With Crises: The Management of Disasters, Riots and Terrorism*, Springfield IL: Charles C. Thomas.

Rosenthal, U., 't Hart, P., Kouzmin. A., and Jarman, A.(1989) "From case studies to theory and recommendations: a concluding analysis," in U. Rosenthal, M. Charles, and P. 't Hart (eds) *Coping with Crises: The Management of Disasters, Riots and Terrorism*, Springfield IL: Charles C. Thomas.

Rosenthal, U., 't Hart, P., Van Duin, M., Boiin, A., Kroon, M., Otten, M., and Overdijk, W. (1994) *Complexity in Urban Crisis Management: Amsterdam's Response to the Bijlmer Air Disaster*, London: James and James.

Rosset, C. (1979) *L'object Singulier*, Paris: Les Editions de Minui (in French).

Rossi, I. (1994) *Community Reconstruction After an Earthquake: Dialectical Sociology in Action*, Westport CT: Praeger.

Sachs, W. (1990) "Development," in W. Sachs (ed.) *The Development Dictionary*, London: Zed Books.

Sahlins, M. (1976) *Culture and Practical Reason*, Chicago: University of Chicago Press.

Saunders, S. and Kreps, G. (1987) "The life history of emergent organization in disaster," *Journal of Applied Behavioral Science* 23: 443–462.

Saylor, A. (1993) *Children and Death*, New York: Plenum Press.

Scanlon, J. (1988) "Winners and losers: some thoughts about the political economy of disasters," *International Journal of Mass Emergencies and Disasters* 6: 47–63.

—— (1997) "Gender and disasters: a second look," *Natural Hazards Observer* 21: 1–2.

Schacter, D. (1996) *Searching for Memory*, New York: Basic Books.

Schmink, M. and Wood, C. (1987) "The political ecology of Amazonia," in P. Little and M. Horowitz (eds) *Lands at Risk in the Third World*, Boulder CO: Westview Press.

Schorr, J. (1987) "Some contributions German Katastrophensoziologie can make to the sociology of disaster," *International Journal of Mass Emergencies and Disasters* 5: 115–135.

Schrag, C. (1967) "Elements of theoretical analysis in sociology," in L. Gross (ed.) *Sociological Theory*, New York: Harper and Row.

Schutz, A. (1967) *Collected Papers Volume 1: The Problem of Social Reality*, Hague: Martinus Nijhoff.

Schwab, W. (1993) "Recent empirical and theoretical developments in sociological human ecology," *Urban Sociology in Transition, Research in Urban Sociology* 3: 29–57.

Schwartz, M. and Paul, S. (1992) "Resource mobilization versus the mobilization of people: why consensus movements cannot be instruments of social change," in A. Morris and C. Mueller (eds) *Frontiers in Social Movement Theory*, New Haven CT: Yale University Press.

Seeck, E. (1980) *Gesetz uber den Katastrophenschutz in Schlesig-Holstein (LkatSG) vom 9 Dezember 1974*, Wiesbaden, Germany: Kommunal und Schul-Verlag A. Heinig (in German).

Seeman, M. (1983) "Alienation motifs in contemporary theorizing," *Social Psychological Quarterly* 46: 171–184.

Sen, A. (1981) *Poverty and Famines: An Essay on Entitlement and Deprivation*, Oxford: Clarendon Press.

Shapin, S. (1995) "Here and everywhere: sociology of scientific knowledge," *Annual Review of Sociology* 21: 289–321.

Shcherbak, I. (1989) *Chernobyl: A Documentary Story*, New York: St. Martin's Press.

Shkilnyk, A. (1985) *A Poison Stronger Than Love: The Destruction of an Ojibwa Community*, New Haven: Yale University Press.

Short. D. (1989) "On defining, describing and explaining elephants (and reactions to them): hazards, disasters, and risk analysis," *International Journal of Mass Emergencies and Disasters* 7: 397–418.

Shrader-Frechette, K. (1991) *Risk and Rationality: Philosophical Foundations for Populist Reforms*, Berkeley: University of California Press.

Shrivastava, P. (1987) "Preventing industrial crises: the challenge of Bhopal," *International Journal of Mass Emergencies and Disasters* 5: 199–221.

Shrivastava, P., Mitroff, I., Miller, D., and Miglani, K. (1988) "Understanding industrial crises," *Journal of Management Studies* 25: 283–303.

Sjoberg, G. (1962) "Disasters and social change," in G. Baker and D. Chapman (eds) *Man and Society in Disaster*, New York: Basic Books.

Sjoberg, G. and Nett, R. (1967) *A Methodology for Social Research*, New York: Harper and Row.

Sloterdjik, P. (1987) *Critique of Cynical Reason*, Minneapolis: University of Minnesota Press.
Smart, B. (1993) *Postmodernity*, London: Routledge.
Smelser, N. (1991) "The social sciences in a changing world society," *American Behavioral Scientist* 34: 518–529.
Smith, K. (1992) *Environmental Hazards: Assessing Risk and Reducing Disaster*, London: Routledge.
Smith, R. (1991) "The inapplicability principle: what *chaos* means for social science," unpublished paper.
Sorokin, P. (1928) *Contemporary Sociological Theories*, NewYork: Harper.
—— (1942) *Man and Society in Calamity: The Effects of War, Revolution, Famine, Pestilence Upon Human Mind*, New York: Dutton.
—— (1943) *Sociocultural Causality, Space, Time: A Study of Referential Principles of Sociology and Social Science*, New York: Russell and Russell.
Stallings, R. (1978) "The structural patterns of four types of organizations in disasters," in E. Quarantelli (ed.) *Disasters: Theory and Research*, Beverly Hills: Sage.
—— (1988) "Conflict in natural disaster," *Social Science Quarterly* 69: 569–586.
—— (1991) "Disasters as social problems? A dissenting view," *International Journal of Mass Emergencies and Disasters* 9: 90 95.
—— (1995) *Promoting Risk: Constructing the Earthquake Threat*, New York: Aldine de Gruyter.
Starosolszky, O. and Melder O. (eds) (1989) *Hydrology of Disasters: Proceedings of the Technical Conference in Geneva, November 1988*, London: James and James.
Stoddard, E. (1968) *Conceptual Models of Human Behavior in Disaster*, El Paso TX: Texas Western Press.
Suren, E. (1982) *Einsatzhinweise bei Massenunfallen und Katastrophen*, Eriauterungen zum Merkblatt, Niedersachsisches: Arzteblatt Nr. 4 (in German).
Susman, P., O'Keefe, P., and Wisner, B. (1983) "Global disasters: a radical approach," in K. Hewitt (ed.) *Interpretations of Calamity: From the Viewpoint of Human Ecology*, London: Allen and Unwin.
Taylor, P. (1992) *War and the Media: Propaganda and Persuasion in the Gulf War*, Manchester: Manchester University Press.
Taylor, V. (1978) "Future directions for research," in E. Quarantelli (ed.) *Disasters Theory and Research*, London: Sage.
Theys, J. and Fabiani, J. (1987) *La Société Vulnérable*, Paris: Presses de l'Ecole Normale Superieure (in French).
Thomas, W. and Thomas, D. (1928) *The Child in America: Behavior Problems and Programs*, New York: Knopf.
Thompson, J. (1967) *Organizations in Action: Social Science Bases of Administrative Theory*, New York: McGraw-Hill.
Thompson, M., Ellis, R., and Wildavsky, A. (1990) *Cultural Theory*, Boulder CO: Westview Press.
Tierney, K. (1980) *A Primer for Preparedness for Acute Chemical Emergencies*, Newark DE: Disaster Research Center, University of Delaware.
—— (1989) "Improving theory and research on hazard mitigation: political economy and organizational perspectives," *International Journal of Mass Emergencies and Disasters* 7: 367–396.

—— (1994) "Property damage and violence: a collective behavior analysis," in M. Bladassare (ed.) *The Los Angeles Riots: Lessons For the Urban Future*, Greenwich CT: JAI Press.

—— (forthcoming) "Toward a critical sociology of risk," *Sociological Forum.*

Tierney, K. and Taylor, V. (1977) "EMS delivery in mass emergencies: preliminary research findings," *Mass Emergencies* 2: 151–158.

Timmerman, P. (1981) *Vulnerability, Resilience and the Collapse of Society: a Review of Models and Possible Climatic Applications*, Toronto: Institute of Environmental Studies, University of Toronto.

Tiryakian, E. (1994) "The new worlds and sociology: an overview," *International Sociology* 9: 131–148.

Toft, B. and Reynolds, S. (1994) *Learning From Disasters*, Oxford: Butterworth-Heinemann.

Torry, W. (1978) "Natural disasters, social structure and changes in traditional societies," *Journal of Asian and African Studies* 13: 167–183.

—— (1979) "Anthropological studies in hazardous environments: past trends and new horizons," *Current Anthropology* 20: 517–541.

—— (1986) "Morality and harm: Hindu peasant adjustments to famines," *Social Science Information* 12: 1–10.

Touraine, A. (1995) *Critique of Modernity*, Oxford: Blackwell.

Towfighi, P. (1991) "Integrated planning for natural and technological disasters," in A. Kreimer and M. Munasinghe (eds). *Managing Natural Disasters and the Environment*, Washington DC: Environment Department, World Bank.

Turing, A. (1936) "On computable numbers with an application to the entscheidungs problem," in *Proceedings of London Mathematical Society*, London: Cambridge.

Turner, B. (1978) *Man-Made Disasters*, London: Wykeham.

—— (1979) "The social etiology of disasters," *Disasters* 3: 53–59.

Turner, B. and Pidgeon, N. (1997) *Man-Made Disasters*, 2nd edition, Oxford: Butterworth-Heinemann.

Turner, R. (1964) "Collective behavior," in R. Faris (ed.) *Handbook of Sociology*, Chicago: Rand McNally.

—— (1989) "Taxonomy as an approach to theory development," *International Journal of Mass Emergencies and Disasters* 7: 265–276.

UNDRO (1982) *Natural Disasters and Vulnerability Analysis*, Geneva: Office of United Nations Disaster Relief Coordinator.

United States General Accounting Office (1997) *Financial Crisis Management: Four Financial Crises in the 1980s*, Washington DC: US General Accounting Office.

Vaughan, D. (1996) *The* Challenger *Launch Decision: Risky Technology, Culture and Deviance at NASA*, Chicago IL: University of Chicago Press.

Vaughan, M. (1987) *The Story of an African Famine: Gender and Famine in Twentieth Century Malawi*, Cambridge: Cambridge University Press.

Von Bretzel, P. and Nagasawa, R. (1977) *Logic, Theory and Confirmation in Sociology*, Washington DC: University Press of America.

Vyner, J. (1988) *Invisible Trauma*, Lexington MA: Lexington Books.

Waddell, E. (1975) "How the Enga cope with frost: responses to climatic perturbations in the central highlands of New Guinea," *Human Ecology* 3: 249–273.

—— (1977) "The hazards of scientism: a review article," *Human Ecology* 5: 67–76.

—— (1983) "Coping with frosts, governments and disaster experts: some reflections based on a New Guinea experience and a perusal of the relevant literature," in K. Hewitt (ed.) *Interpretations of Calamity: From the Viewpoint of Human Ecology*, London: Allen and Unwin.

Waldrop, M. (1992) *Complexity: The Emerging Science at the Edge of Order and Chaos*, New York: Simon and Schuster.

Walford, C. [1879] (1980) *Famines of the World: Past and Present*, New York: Burt Franklin.

Wallace, A. (1956a) *Human Behavior in Extreme Situations*. Washington DC: National Academy of Sciences.

—— (1956b) *Tornado in Worcester*, Washington DC: Committee on Disaster Studies, National Academy of Sciences.

Wallace, W. (1991) "Standardizing basic sociological concepts," *Perspective* 14: 17–34.

Walker, P. (1989) *Famine Early Warning System: Victims and Destitution*, London: Earthscan.

Warheit, G. (1968) "The impact of major emergencies on the functional integraton of four American communities," PhD dissertation, Ohio State University.

—— (1996) *The Fire Department in Disaster Operations in the 1960s*, Newark DE: Disaster Research Center, University of Delaware.

Warren, K. (ed.) (1994) *Ecological Feminism*, London: Routledge.

Watts, M. (1983a) *Silent Violence: Food, Famine and Peasantry in Northern Nigeria*, Berkeley: University of California Press.

—— (1983b) "On the poverty of theory: natural hazards research in context," in K. Hewitt (ed.) *Interpretations of Calamity: From the Viewpoint of Human Ecology*, London: Allen and Unwin.

Watts, M. and Bohle, H. (1993) "The space of vulnerability: the causal structure of hunger and famine," *Progress in Human Geography* 17: 143–167.

Waugh, W. (1997) "The fiscal risk of all-hazards emergency management *or* the political hazards in rational policy," *International Journal of Public Administration*: 1–23.

Weber, M. (1947) *Theory of Social and Economic Organization*, New York: Free Press.

—— (1958) *Politics as a vocation*," in H. Gerth and C.Mills (eds) *From Max Weber: Essays in Sociology*, New York: Oxford University Press.

—— (1978) *Economy and Theory: An Outline of Interpretive Sociology* Volume 2, Berkeley CA: University of California Press.

Weick, K. (1990) "The vulnerable system: an analysis of the Tenerife air disaster," *Journal of Management* 6: 571–593.

—— (1993) "The collapse of sensemaking in organizations: the Mann Gulch disaster," *Administrative Science Quarterly* 38: 628–652.

Weller, J. (1972) "Innovations in anticipation of crisis: organizational preparations for natural disasters and civil disturbances," unpublished PhD dissertation, Ohio State University.

—— (1974) *Organizational Innovation in Anticipation of Crisis*, Newark DE: Disaster Research Center, University of Delaware.

Weller, J. and Quarantelli, E. (1973) "Neglected characteristics of collective behavior," *American Journal of Sociology* 79: 665–685.

Wenger, D. (1973) "The reluctant army: the functioning of police departments during civil disturbance," *American Behavioral Scientist* 16: 312–325.

—— (1978) "Community response to disaster," in E. Quarantelli (ed.) *Disasters: Theory and Research*, Beverly Hills CA: Sage.

Wenger, D. and Weller, J. (1973) "Disaster subcultures: the cultural residue of community disasters," Preliminary Paper # 9, Newark DE: Disaster Research Center, University of Delaware.

Westgate, K. and O'Keefe, P. (1976) *Some Definitions of Disaster*, Bradford, England: Disaster Research Institute, University of Bradford.

Whimster, S. and Lash, S. (eds) (1987) *Max Weber, Rationality and Modernity*, London: Allen and Unwin.

White, G. (1964) "The choice of use in resource management," *Natural Resources Journal* 1: 25–40.

Whyte, A. and Burton, I. (ed.) (1980) *Environmental Risk Assessment*, New York: Wiley.

Wijkman, A. and Timberlake, L. (1984) *Natural Disasters: Acts of God or Acts of Man?*, London: Earthscan.

Williams, H. (1954) "Fewer disasters, better studied," *Journal of Social Issues* 10: 5–11.

Winchester, P. (1992) *Power, Choice and Vulnerability: A Case Study of Disaster Mismanagement in South India 1977–1988*, London: James and James.

Wisner, B. (1988) *Power and Need in Africa*, London: Earthscan.

—— (1990) "World launches international decade for natural disaster reduction," *UNDRO News* (Special Issue, Jan/Feb).

—— (1993) "Disaster vulnerability, geographical scale and existential reality," in H. Bohle (ed.) *Worlds of Pain and Hunger: Geographical Perspectives on Disaster Vulnerability and Food Security*, Fort Lauderdale: Freiburg Studies in Development Geography.

Wright, J. and Rossi, P. (eds) (1981) *Social Science and Natural Hazards*, Cambridge MA: Abt Books.

Wright, K., Ursano, R., Bartone, P., and Ingraham, L. (1990) "The shared experience of catastrophe: an expanded classification of the disaster community," *American Journal of Orthopsychiatric* 60: 35–42.

Wright, W. (1992) *Wild Knowledge: Science, Language, and Social Life in a Fragile Environment*, Minneapolis MN: University of Minnesota Press.

Young, M. (1988) *The Metronomic Society: Natural Rhythms and Human Timetables*, Cambridge MA: Harvard University Press.

Young, M. and Schuller, T. (ed.) (1988) *The Rhythms of Society*, New York: Routledge.

Zaman, M. (1994) "Ethnography of disasters: making sense of flood and erosion in Bangladesh," *Eastern Anthropologist* 47: 129–155.

Zeigler, D., Johnson, J., and Brunn, S. (1983) *Technological Hazards*, Washington DC: Association of American Geographers.

Zerubavel, E. (1981) *Hidden Rhythms: Schedules and Calendars in Social Life*, Chicago: University of Chicago Press.

Zimmerman, M. (1994) *Contesting Earth's Future: Radical Ecology and Postmodernity*, Berkeley CA: University of California.

Zimmermann, E. (1983) *Political Violence, Crises and Revolutions: Theories and Research*, Boston: Hall.

Zurcher, L. (1968) "Socio-psychological functions of ephemeral roles: a disaster work crew," *Human Organization* 27: 281–297.

INDEX